ENERGY 2020

The Tenth International Conference on Smart Grids, Green Communications and IT Energy-aware Technologies

September 27th – October 1st, 2020

ENERGY 2020 Editors

Vivian Sultan, California State University Los Angeles, USA

Michael Kuhn, University of Hamburg, Germany

Eric MSP Veith, OFFIS e.V. - Oldenburg, Germany

Daniele Jahier Pagliari, Politecnico di Torino, Italy

Massimo Poncino, Politecnico di Torino, Italy

Printed from e-media with permission by:

Curran Associates, Inc.
57 Morehouse Lane
Red Hook, NY 12571

Some format issues inherent in the e-media version may also appear in this print version.

International Academy, Research, and Industry Association (IARIA)
412 Derby Way
Wilmington, DE 19810

Phone: (408) 893-6407
Fax: (408) 527-6351

petre@iaria.org

Additional copies of this publication are available from:

Curran Associates, Inc.
57 Morehouse Lane
Red Hook, NY 12571 USA
Phone: 845-758-0400
Fax: 845-758-2633
Email: curran@proceedings.com
Web: www.proceedings.com

Table of Contents

Load and Demand Side Flexibility Forecasting 1
Rocio Alvares Merce, Etta Grover-Silva, and Johanna Le Conte

Predictive Model for Power Outages 7
Vivian Sultan

A Framework for Strategy Selection of Atomic Entities in the Holonic Smart Grid 11
Alexander Wallis, Sascha Hauke, Rolf Egert, and Max Muhlhauser

Improving Energy Efficiency of Scientific Data Compression with Decision Trees 17
Michael Kuhn, Julius Plehn, Yevhen Alforov, and Thomas Ludwig

ArduPower v2: Open and Modular Power Measurement for HPC Components 24
Daniel Bremer, Michael Kuhn, and Mohammad Reza Heidari

Large-Scale Co-Simulation of Power Grid and Communication Network Models with Software in the Loop 30
Eric Veith, Jawaid Kazmi, and Stephan Balduin

Intrusion Detection in Smart Grid Distribution Domain Using Deep Ruptures Detection 37
Sarah Chahine and Chafic Mokbel

The Importance of Accurate Battery Models for Power Assessment in Smart Energy Systems 44
Yukai Chen and Sara Vinco

Powertrain Modeling and Range Estimation of The Electric Bus 50
Donkyu Baek

When Privacy Protection Meets Non-Intrusive Load Monitoring: Trade-off Analysis and Privacy Schemes via Residential Energy Storage 56
Sangyoung Park and Andrea Cominola

ENERGY 2020

Forward

The Tenth International Conference on Smart Grids, Green Communications and IT Energy-aware Technologies (ENERGY 2020) continued a series of events considering Green approaches for Smart Grids and IT-aware technologies. It addressed fundamentals, technologies, hardware and software support, applications and challenges.

There is a perceived need for a fundamental transformation in IP communications, energy-aware technologies and the way all energy sources are integrated. This is accelerated by the complexity of smart devices, the need for special interfaces for an easy and remote access, and the new achievements in energy production. Smart Grid technologies promote ways to enhance efficiency and reliability of the electric grid, while addressing increasing demand and incorporating more renewable and distributed electricity generation. The adoption of data centers, penetration of new energy resources, large dissemination of smart sensing and control devices, including smart home, and new vehicular energy approaches demand a new position for distributed communications, energy storage, and integration of various sources of energy.

We take here the opportunity to warmly thank all the members of the ENERGY 2020 technical program committee, as well as all the reviewers. The creation of such a high quality conference program would not have been possible without their involvement. We also kindly thank all the authors who dedicated much of their time and effort to contribute to ENERGY 2020. We truly believe that, thanks to all these efforts, the final conference program consisted of top quality contributions. We also thank the members of the ENERGY 2020 organizing committee for their help in handling the logistics of this event.

ENERGY 2020 Chairs

ENERGY 2020 Steering Committee
Eric MSP Veith, OFFIS e.V. – Oldenburg, Germany
Michael Negnevitsky, University of Tasmania, Australia
Dragan Obradovic, Siemens - Corporate Technology, Munich, Germany
Vivian Sultan, California State University Los Angeles, USA
Mark Apperley, University of Waikato, New Zealand
Steffen Fries, Siemens, Germany

ENERGY 2020 Publicity Chair
Joseyda Jaqueline More, Universitat Politecnica de Valencia, Spain
Marta Botella-Campos, Universitat Politecnica de Valencia, Spain

ENERGY 2020 Industry/Research Advisory Committee
Hongbo Sun, Mitsubishi Electric Research Laboratories, USA
Daisuke Mashima, Advanced Digital Sciences Center, Singapore

Sambaran Bandyopadhyay, IBM Research, India
Chresten Træholt, Center for Electric Power and Energy - CEE | Technical University of Denmark - DTU, Denmark
Vincenzo Gulisano, Chalmers University of Technology, Sweden
Marco Lützenberger, Technische Universität Berlin, Germany
Evangelos Pournaras, ETH Zurich, Switzerland
Chun-Hsi Huang, University of Connecticut, USA
Magnus Almgren, Chalmers University of Technology, Sweden

ENERGY 2020
Committee

ENERGY 2020 Steering Committee

Eric MSP Veith, OFFIS e.V. – Oldenburg, Germany
Michael Negnevitsky, University of Tasmania, Australia
Dragan Obradovic, Siemens - Corporate Technology, Munich, Germany
Vivian Sultan, California State University Los Angeles, USA
Mark Apperley, University of Waikato, New Zealand
Steffen Fries, Siemens, Germany

ENERGY 2020 Publicity Chair

Joseyda Jaqueline More, Universitat Politecnica de Valencia, Spain
Marta Botella-Campos, Universitat Politecnica de Valencia, Spain

ENERGY 2020 Industry/Research Advisory Committee

Hongbo Sun, Mitsubishi Electric Research Laboratories, USA
Daisuke Mashima, Advanced Digital Sciences Center, Singapore
Sambaran Bandyopadhyay, IBM Research, India
Chresten Træholt, Center for Electric Power and Energy - CEE | Technical University of Denmark - DTU, Denmark
Vincenzo Gulisano, Chalmers University of Technology, Sweden
Marco Lützenberger, Technische Universität Berlin, Germany
Evangelos Pournaras, ETH Zurich, Switzerland
Chun-Hsi Huang, University of Connecticut, USA
Magnus Almgren, Chalmers University of Technology, Sweden

ENERGY 2020 Technical Program Committee

Santosh Aditham, University of South Florida, USA
Mohammad Pourmahmood Aghababa, Urmia University of Technology, Iran
Kamal Al-Haddad, École de technologie supérieure, Montreal, Canada
Miltos Alamaniotis, University of Texas at San Antonio, USA
Magnus Almgren, Chalmers University of Technology, Sweden
Mark Apperley, University of Waikato, New Zealand
Navid Bayati, Aalborg University, Denmark
Andreas Berl, Technische Hochschule Deggendorf, Germany
Vito Calderaro, University of Salerno, Italy
M. Girish Chandra, TCS Research & Innovation, India
Dana-Alexandra Ciupageanu, University Politehnica of Bucharest, Romania
Daniele Codetta, University of Piemonte Orientale, Italy
Luigi Costanzo, Università degli Studi della Campania Luigi Vanvitelli, Italy
Margot Deruyck, Universiteit Gent - IMEC - WAVES, Belgium
Jiajun Duan, GEIRINA Inc, USA
Niklas Ebell, Friedrich-Alexander-University Erlangen-Nürnberg, Germany
Rolf Egert, Technische Universität Darmstadt, Germany
Kevin Ellett, Indiana University, USA
Meisam Farrokhifar, University of Groningen, The Netherlands

Mahmoud Fotuhi-Firuzabad, Sharif University of Technology, Tehran, Iran
Steffen Fries, Siemens, Germany
Vincenzo Galdi, University of Salerno, Italy
Luis Garcia, USC ISI, USA
Anandha Gopalan, Imperial College London, UK
Saman K. Halgamuge, The University of Melbourne, Australia
Yunzhi Huang, Pacific Northwest National Laboratory - U.S. Department of Energy, USA
Philip Johnson, University of Hawaii, USA
Hadis Karimipour, University of Guelph, Canada
Christoph Klemenjak, University of Klagenfurt | Institute of Networked and Embedded Systems, Austria
Tobias Küster, Technische Universität Berlin (TU Berlin), Germany
Duc Van Le, Nanyang Technological University, Singapore
Gerard Ledwich, Queensland University of Technology, Australia
Shunbo Lei, University of Michigan, USA
Yiu-Wing Leung, Hong Kong Baptist University, Kowloon Tong, Hong Kong
Tek Tjing Lie, Auckland University Of Technology, New Zealand
Zhenhua Liu, Stony Brook University (SUNY at Stony Brook), USA
Rafael Mayo-García, CIEMAT, Spain
Amin Mokari, School of Electrical Engineering & Robotics | Queensland University of Technology, Australia
Hugo Morais, Universidade de Lisboa, Portugal
Fabio Mottola, University of Naples Federico II, Italy
Gero Mühl, Universitaet Rostock, Germany
Hamidreza Nazaripouya, University of California, Riverside, USA
Michael Negnevitsky, University of Tasmania, Australia
Claudiu Oprea, Technical University of Cluj-Napoca, Romania
Youssef Ounejjar, ETS, Montreal, Canada
Daniele Jahier Pagliari, Politecnico di Torino, Italy
Shalini Pandey, University of Minnesota, USA
Thanasis Papaioannou, Athens University of Economics and Business (AUEB), Greece
Marco Pasetti, University of Brescia, Italy
Nilavra Pathak, University of Maryland Baltimore County, USA
Faz Rahman, The University of New South Wales, Sydney, Australia
Djamila Rekioua, University of Bejaia, Algeria
Jan Richling, South Westphalia University of Applied Sciences, Germany
Angela Russo, Politecnico di Torino, Italy
Farhad Shahnia, Murdoch University, Australia
Vivian Sultan, California State University Los Angeles, USA
Hongbo Sun, Mitsubishi Electric Research Laboratories, USA
Jay Taneja, University of Massachusetts, Amherst, USA
François Vallee, University of Mons, Belgium
Eric MSP Veith, OFFIS e.V. - Oldenburg, Germany
Alekhya Velagapudi, University of Pittsburgh's School of Computing and Information, USA
Jian Xu, Texas Reliability Entity (Texas RE), USA
Sean Yaw, Montana State University, USA
Technical Co-Sponsors and Logistic Supporters

Copyright Information

Load and Demand Side Flexibility Forecasting

Rocio Alvarez Merce
Ecole Polytechnique
Palaiseau Cedex, France
Email: rocio.alvarez-merce@ecoco2.com

Etta Grover-Silva
Eco CO2
Nanterre, France
Email: etta.grover-silva@ecoco2.com

Johanna Le Conte
Eco CO2
Nanterre, France
Email: johanna.leconte@ecoco2.com

Abstract—**Recent developments in energy metering technologies have allowed electric load data to be more easily accessible. Services that use this data to inform customers can raise awareness about electricity consumption and provide suggestions to encourage energy efficient behavior. Quantifying the flexibility of the demand combined with accurate predictions of the total electric load allow information services to provide suggestions to end-users on how to reduce electric consumption that are appliance and time specific. With the arrival of new electric generation technologies, such as photovoltaic or wind energy, demand side flexibility will play an important role in the optimization of the future electric system. The accurate prediction of demand flexibility at a building level, therefore can contribute to the optimization of load scheduling. This study presents an effective multi-step technique to forecast the average hourly demand flexibility for a household, using neural networks, unsupervised clustering techniques and an optimization algorithm, combined with statistical studies. The model is mainly based on the historical electric use of a building and does not require significant computational capacity, thus making it widely applicable and informative for residential customers, helping them improve their behavior to be more energy efficient in the future.**

Keywords–non-intrusive load monitoring; demand flexibility; long short-term memory; recurrent neural network.

I. INTRODUCTION

Worldwide, the energy sector is increasing the penetration of decentralized renewable power generation systems and therefore, reducing more traditional centralized fossil fuel generation. This transition presents several challenges, such as moving from a more stable and controllable power generation to a more volatile and less predictable one. Mitigating this volatility and simultaneously decreasing the number of conventional power generators makes it harder to balance out supply and demand in order to ensure a stable and reliable grid. In addition, emissions from buildings have risen in the last few years, reaching an all-time high in 2018 [1], making this transition even more challenging. In particular, households in the European Union (EU) represented one-fourth of its total final energy consumption [2]. This results from several factors, including extreme weather conditions that increase energy demand for heating and cooling and inefficient behavioral habits that result in high energy use which is unnecessary for the comfort of building inhabitants.

Accordingly, demand side management is important to reduce overall emissions of buildings while guaranteeing end-user comfort. Accurate forecasts of individual building electric loads are crucial to effectively inform end-users about their energy consumption habits. The willingness of end-users to change their energy use habits can therefore provide demand flexibility. Accurate predictions of the energy demand and demand flexibility at a building level can help encourage energy efficient behavior, stabilize the electricity grid and reduce the electricity bill, while accelerating the sustainable energy transition.

Recent developments in technology have allowed energy consumption data to be more easily accessible. Several studies show that real-time feedback about appliance specific enery use in energy efficiency awareness programs results in the highest energy savings [3]. In this regard, an awareness raising project has been put in place by the company Eco CO2 in the context of a public tender in France put forth by ADEME related to a funding mechanism called Investissement d'avenir [4]. This awareness raising service called Tableu de Bord de l'Habitat (TBH) Alliance aims to test different interfaces with varying information about the electric load of each user to encourage energy efficient behaviour. A digital tablet and a website is available for users to observe their electric load data, load profile analysis metrics and suggestions to reduce their consumption. In this context, quantifying demand flexibility combined with accurate predictions of the total electric load can lead to services that provide action plans to reduce energy consumption that are appliance and time specific.

Models used in building energy analysis can be grouped into two categories: the top-down approach, and the bottom-up approach [5]. Top-down approaches do not consider the occupant behavior inside, and very little information is needed about the building, they rely mostly on historical consumption data. Bottom-up techniques take into consideration the physical characteristics of a building and occupant behavior resulting in more detailed models. The bottom up approaches require a significant amount of detailed input information about each building and often require complex models that require a high computational time. Therefore, a pure bottom-up approach is not feasible for the assessment of a large group of end-users where detailed information is not available.

Many different top-down methods have been developed for load forecasting, such as curve fitting using numerical methods or machine learning algorithms [6]–[9]. Load forecasting is a complex multi-step time series regression problem. Some forecasting techniques, such as curve fitting using numerical methods, do not provide accurate results as they often fail to track seasonality and trends accurately. Machine learning algorithms, such as neural networks are more effective at integrating seasonal trends. In particular, they can be applied to energy consumption data to forecast electric load profiles. In spite of new developments in literature and applied modeling, this remains to be a difficult problem [6].

Traditional machine learning algorithms are often ineffective at predicting sequential data, where each data point represents an observation at a certain point in time. The algorithm assumes that the data is non-sequential, and that each data point is independent of the others. As a result, the inputs are analyzed independently, without the intrinsic temporal dependencies. Consequently, some models are successful at predicting a single value, but they fail to attain multi-step forecasting. A benefit of neural network models over many other machine learning techniques is that they are able to compute multi-step predictions. This is useful in time series forecasting as the forecasts are multiple future time steps. In the field of building energy consumption forecasting methods, Artificial Neural Networks (ANN) are the most commonly used models for making load and energy use predictions [7]. ANN modeling techniques have been used to estimate energy consumption in multiple studies [6], [8], [10], [11] using Convolutional, Nonlinear Autoregressive and Feed Forward Neural Networks.

Recurrent Neural Network (RNN) has been proven to be a powerful tool for modeling sequential data as a regression time-series problem. The RNN is able to remember the analysis that was done up to a given point by maintaining a state, considering past observations. The state can be thought of as the memory of the RNN, which captures information about whats been previously calculated and is integrated into each step in the training process. This allows RNNs to process information sequentially and exhibit temporal behavior for a time sequence while retaining information from the past. Nonetheless, there are a few challenges in the effective implementation of this algorithm. Recurrent networks are computationally intensive since they keep track of past states. Some other common issues that may appear during the training phase are the vanishing or exploding gradient. As a result, the simple Vanilla RNN is not useful for long sequences of data. To solve these problems, a specific type of RNN that maintains a strong gradient over many time steps is used, thus being able to efficiently work with long sequences: the Long Short-Term Memory (LSTM) model.

Once accurate predictions of the load profile are acquired, it is necessary to then assess the flexibility of this future load. Several strategies to quantify the demand response of the residential sector to optimize electricity consumption are present in literature. Furthermore, the individual flexibility of different smart appliances has been quantified. Sancho Tomas et al. [12] applied a time-inhomogeneous Markov process to model energy variations over time for different appliances. Laicane et al. [13] investigated the potential for demand side management to reduce peak load. Load shifting algorithms were developed by Dlamini and Cromieres [14] to achieve significant household load reduction. DHulst et al. [15] presented a demand response flexibility analysis based on measurements from appliances within a large-scale implementation of smart grid technologies in a distribution grids project. These studies allow for the quantification of the flexibility of individual appliances.

In this paper, a LSTM model is used to predict the load profile of individual households. This predicted load profile is then decomposed into specific load categories using a combination of both top-down and bottom-up non-intrusive analysis. The prediction of appliance specific load profiles then allows for the forecast of total demand response potential of a household and the associated flexibility. Firstly, limited information about households is gathered through a questionnaire about the building energy systems and operational set points. Then, an effective supervised learning algorithm is used to forecast the energy consumption of the households, using historical load profile data. Thirdly, in a top-down approach, a decomposition algorithm is used to partition the predicted load profile into two main categories: active (manual control of appliances by inhabitants) and inactive (appliances that cycle automatically and are not controlled directly by inhabitants) loads. These categories are then further separated by applying statistical studies of typical appliance uses to provide an estimate of the expected energy use per device [16]. Finally, an optimization algorithm is used to reconstruct a load forecast and the average flexibility of the demand is determined for each hour of the day. This multi-step hybrid approach is applied to a case study of three residential clients and the results are presented. The following sections of this paper are composed of Section II describing the methodology of the algorithm including Section II-A detailing the forecasting algorithm, Section II-B describing the estimation of the load flexibility and Section II-C presenting details on the evaluation of the load predictions. It is followed by Section III, which describes the case study where results are presented in Section IV and final conclusion in Section V.

II. Methodology

The electric load data used for this study was collected through the services offered by the company Eco CO2. Using a LSTM model, total load profiles were forecasted for each household. These forecasted load profiles are then analyzed with a non-intrusive load decomposition technique. Additionally, an optimization algorithm is used to reconstruct the hourly load profile for each appliance type. Finally, the hourly flexibility potential for each household is determined.

A. *Forecasting*

For the load forecasting, an LSTM model was used. The RNN has two hidden layers: a 200-neuron LSTM layer, and a 100-neuron dense layer, implemented in python with the TensorFlow and Keras sequential model packages. To determine the best choice of hyperparameters for the model, a sensitivity test was performed on multiple hyperparameters of the LSTM algorithm. Different architectures of LSTM networks were compared, each one with different hyperparameters, reaching an optimal compromise between forecasting accuracy and low computational time. The number of time steps used as input was $n_{input} = 24$ prior observations, and both the input and output layers have the same number of time steps ($n_{input} = n_{output}$).

The network was trained on 43 weeks of historical data in each run. This 43 week period was choosen based on the maximum historical data available for all users. This was repeated four times in order to obtain 1 full month of load predictions for each user. The default number of epochs and batch size for all analysis was $n_{epochs} = 400$ and $n_{batch} = N/25$, respectively, being N the total number of training samples.

Finally, the rectifier (ReLU) was chosen as the activation function, and ADAM (Adaptive Moment Estimation [17]) as the optimizer.

B. Energy Flexibility

To determine the flexibility potential of a 24-hours period for a household, a two-step approach was applied to the load forecasts. The load category partitioning algorithm published in [3] is applied to forecasted profiles obtained from the LSTM model to classify the total weekly energy use into eight categories listed in Table I. Active load periods are classified as the time intervals with a relatively high value and high variance.

TABLE I. HOUSEHOLD CHARACTERISTICS

Active load categories	*Inactive load categories*
Cooking	Domestic Hot Water (DHW)
Lighting	Refrigeration
Multimedia active	Multimedia standby
Washing	Continuous Mandatory Ventilation (CMV)

The load categories classified as inactive were assumed to be a constant percentage of the inactive load profile for the whole week period. The load categories classified as active required an optimization algorithm to assign the active consumption to different time periods of the active load profile. Therefore, identifying the expected times of use of each category. This was achieved by minimizing the objective function:

$$min \sum_{t}^{T} X_{act,t} * p_{act} - b_{act,t} \tag{1}$$

where $X_{act,t}$ is the unknown binary variable matrix where each column corresponds to a specific active load category time series, $b_{act,t}$ is a column vector with the forecasted total energy use of each category for the week, and $p_{act}$is a column vector with the average hourly power values for each category. These last values were collected from statistical household energy consumption studies [18]–[21]. In addition, four constraints were defined, one per each active load category:

$$\sum_{t}^{T} j_t \leq j_{tot} \tag{2}$$

where j_t is the value of the active category j corresponding to timestep t and j_{tot} is the weekly energy use obtained in the previous step, corresponding to the same category j of the active consumption.

With this two-step approach, hourly load profiles were obtained for each consumption category. The average hourly flexibility potential was then determined. Flexibility depends on both the total load value and the end use device or category. For categories labeled as inactive, reducing consumption entails changing the temperature set-point for different inactive appliances or reducing the consumption of phantom loads. Literature studies show that Domestic Hot Water (DHW) demand response potential can be reduced 3% by decreasing the set-point by 3°C, while refrigeration can attain a 40% reduction by increasing its set-point by 3°C [15]. Standby or phantom loads from the multimedia category have shown a 17% flexibility potential if devices are unplugged when not in use [22]. For categories labeled as active, demand side flexibility correlates more with postponing the cycle of active appliances: 1 washing event over 3 can be shifted in order to increase flexibility potential [15].

C. Forcasting evaluation

Different criteria can be used to evaluate the performance of the regression forecasting model. Commonly used metrics are the coefficient of determination (R-squared, R^2), the Root Mean Square Error (RMSE), and the Mean Absolute Error (MAE). These error measures are defined in (3), (4) and (5), respectively.

$$R^2 = \frac{\sum_{i=0}^{N}(y_i - \hat{y}_i)^2}{\sum_{i=0}^{N}(y_i - \bar{y}_i)^2} \tag{3}$$

$$RMSE = \sqrt{\frac{\sum_{i=0}^{N}(y_i - \hat{y}_i)^2}{N}} \tag{4}$$

$$MAE = \frac{1}{N}\sum_{i=0}^{N}|y_i - \hat{y}_i| \tag{5}$$

where y_i denotes the observed values from the test set, $\hat{y}_i$ the predicted values, and $\bar{y}$ the mean of the observed data as defined in (6).

$$\bar{y} = \frac{1}{N}\sum_{i=0}^{N} y_i \tag{6}$$

The RMSE is the most used evaluation metric for regression models. On the other hand, averaging values makes MAE more robust to outliers while the RMSE gives a relatively high weight to large errors. For the coefficient of determination, high values are preferable. However, even a model with low R^2 can be accurate if the RMSE is low [23].

III. CASE STUDY

The data used for this study was collected through the services offered by the company Eco CO2. Load data from 3 French households was collected through sensors that are capable of reading and transmitting the total electric load data for each household. Additionally, limited information about the households was gathered through a questionnaire. The characteristics of the case study households is summarized in Table II.

TABLE II. HOUSEHOLD CHARACTERISTICS

	Household 1	*Household 2*	*Household 3*
Surface (m^2)	100	110	160
Heating type	Natural gas	Natural gas	Natural gas
DHW	electric	electric	electric
CMV	yes	no	yes
Cooking	electric	electric	electric
Refrigeration	2	3	3 + freezer
Washing	3	3	3
Multimedia	6	6	7
Annual consumption (kWh)	5786	10193	4172

Household 1 shows an annual electric consumption of 5786 kWh. Household 3 presents the lowest measured value, roughly exceeding 4000 kWh, whereas household 2 uses more than 10000 kWh per year. The maximum consumption attained for one-hour period ranges between 4 kWh to almost 7 kWh, for users 2 and 3, respectively.

All households present gas heating, electric cooking devices and roughly the same number of washing and multimedia appliances. Regarding the refrigeration category, household 1 presents the lowest number of appliances (two). Both households 2 and 3 have three refrigeration devices, taking into consideration that, for the latter, one of these appliances is a freezer. User 3 is the only one without electric DHW and user 2 does not have a Continuous Mandatory Ventilation (CMV) system. For this reason, these end-use appliances will not be considered for these users during the decomposition algorithm. In addition, the studied households present different surface dimensions. This variation is taken into account in the optimization model when applying normalized values of statistical household energy consumption studies. For example, the lightning category is highly dependent on the area of the household.

IV. Results

The LSTM model used to forecast electricity consumption was efficient after finding the optimal hyper-parameters. Error measure results are shown in Table III.

TABLE III. Performance of forecasting model

Household ID	R^2	*RMSE*	*MAE*
1	0.692	0.432	0.334
2	0.625	0.464	0.402
3	0.496	0.157	0.125

R-squared values were calculated for the monthly time series forecasts for each user. The calculated values ranged from 0.496 to 0.692, showing a good fit to the test values. The calculated RMSE and MAE showed low values for all cases. Household 3 showed smaller results not because of under fitting issues, but because its average electricity consumption was lower than the other two cases, hence leading to smaller forecasted values and metric results.

The optimization algorithm was then applied to determine the partitioned consumption forecast as shown in Figure 1. Active and inactive loads are highlighted in shades of gray. An example day for each user is presented in Figure 1 with the associated partitioned load curve for the different categories presented in Table I.

Of the three users analyzed, household 3 did not have electric domestic hot water and household 2 did not have a CMV. Accordingly, during the decomposition algorithm, these end use appliances were not considered for these users. The variability of the total consumption may change according to each user and their energy behavior. However, an overall decrease of global consumption occurs for all studied cases during nighttime hours, around 12am, or between 1am and 3am.

Overall, most of the consumption corresponds to the inactive category, reaching 60% and up to 78% of the total consumption. This is mainly caused by the use of DHW,

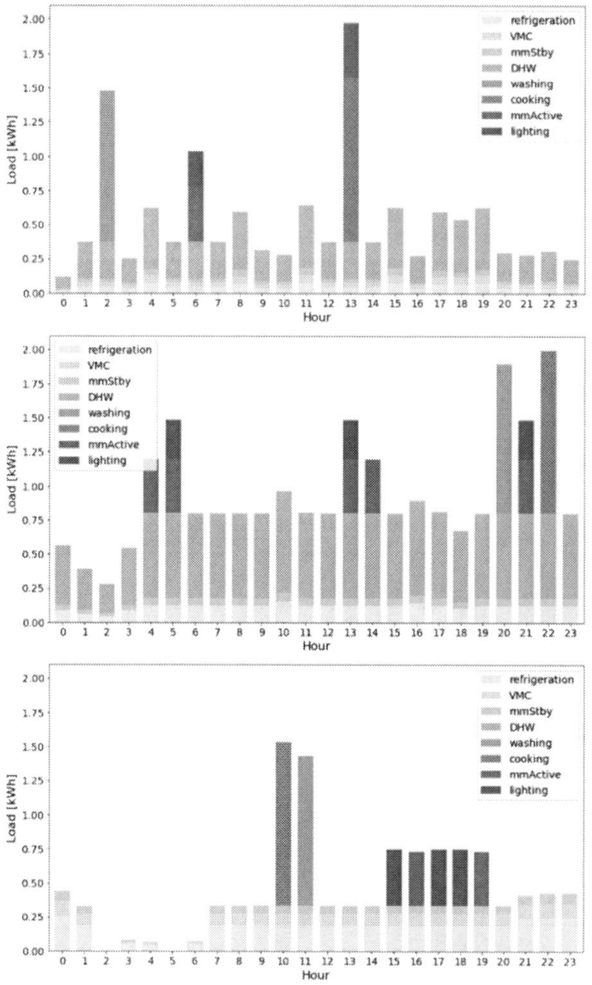

Figure 1. Example electric load forecasts partitioned into appliance categories for a 24-hour period for household 1 (top), household 2 (middle) and household 3 (bottom). Inactive load categories are shown at the bottom of each graph, while active loads are shown at the top of the stacked bars

one of the highest consumption appliances, followed by the refrigeration category. As could be expected, inactive load consumption stays almost constant for all forecasted periods, while active consumption changes depending on user behavior.

The consumption of the active part is dominated by the multimedia and lighting categories, showing consumption peaks several times a day. The washing and cooking categories are only active approximately once every 24 hours. For all end-users, active periods corresponding to multimedia and lightning categories usually occur during the same intervals. These intervals are often between 4am and 6am, or during the afternoon between 1pm and 19pm. With respect to the cooking and washing categories, they occur more often between 10am and 1pm or between 8pm to 10pm.

It is important to notice when these forecasted active and peak inactive periods occur during the day, considering that these could be the possible periods where demand response programs could be more efficiently deployed.

The final average forecasted flexibility potential for each

household is shown in Figure 2. These results are based on the possible consumption savings obtained from the different guidelines mentioned in Section II-B: for inactive loads, reducing consumption entails diminishing the consumption of phantom loads or changing the temperature set point; for active loads, demand side flexibility correlates more with postponing the cycle of active appliances.

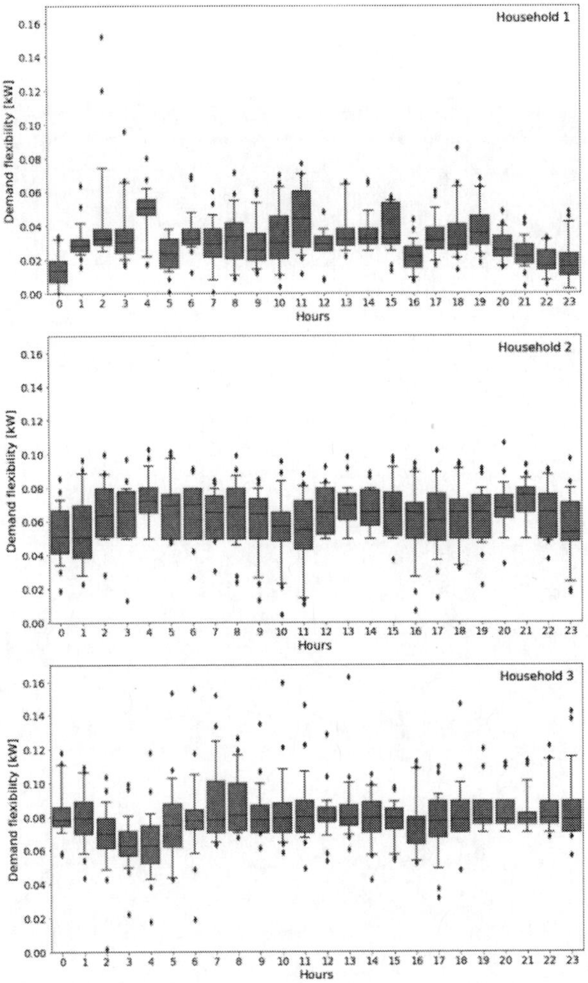

Figure 2. Average demand flexibility values for the three households for a 24-hour period

Average flexibility potential is dependent on the household characteristics, time of day and inhabitant behavior. For all cases, the smallest values for a 24-hour period are found around midnight, between 11pm and 1am, or during the early morning, between 3am and 4am.

Household 1 shows the lowest flexibility values, mostly between 0 and 0.08 kW. This is explained by the characteristics of the household, such as having the lowest number of refrigeration appliances (see Table II) and refrigeration consumption values of all study cases. The adjustment of refrigeration temperature set points results in the largest flexibility in the inactive category, thus not allowing for significant load flexibility in this particular case. Moreover, flexibility associated with changing domestic hot water temperature set points result in only 3% savings. Therefore, for this household, where 70% of the inactive load corresponds to the use of DHW, this results in overall lower flexibility potential.

On the contrary, household 3 presents the highest flexibility values, around 0.08 kW for every hour. This household has the lowest total electricity consumption of all three and does not have electric DHW, usually responsible of a large percentage of the inactive loads. Additionally, more than 50% of his inactive consumption corresponds to the refrigeration category, resulting in a possible 40% inactive load flexibility.

Finally, household 2 shows the greatest variability in the flexibility potential values since it is the household with the highest total consumption. Besides, during all the studied period, the percentage of the active load is higher and more variable, providing a higher active load flexibility but a lower inactive load flexibility.

V. Conclusion

This paper has presented an effective multi-step technique to forecast the average hourly demand flexibility of a household. This model is widely applicable, does not require high computational capacity, and is also compatible with the type and resolution of data available through the massive deployment of smart meters. This solution allows for end-users to learn about their energy use and receive behavior adjustment suggestions for future possible use to encourage energy efficient behavior in advance.

The average demand flexibility values vary between 0.015kW and 0.08kW for each hour depending on the household characteristics, time of day and user behavior. For all cases, the smallest values for a 24-hour period are found around midnight, between 11pm and 1am, or during the early morning, between 3am and 4 am, where it has been noted that the load is relatively low.

The largest flexibility comes from the inactive part of the consumption: changes of temperature set points and unplugging unused multimedia devices causing phantom loads that increase total consumption. Particularly, the adjustment of refrigeration set points results in the largest flexibility values in the inactive category, while DHW usage leads to smallest values: domestic hot water appliances are often responsible of a large percentage of the inactive consumption but its temperature set point change results in only 3% savings. Active consumption appliances contribute to smaller flexibility values since it concerns end-user active behavior, rescheduling activities, postponing appliances or actively changing consumption habits.

Overall, the individual demand flexibility of each user is limited, between 15W and 80W every hour, but the aggregate demand flexibility is interesting to exploit in a massive deployment program. With the arrival of new electric generation technologies, such as photovoltaic or wind energy, demand side flexibility will play an important role in the optimization of the future electric system.

References

[1] "Buildings sector CO2 emissions in the sustainable development scenario," International Energy Agency, 2019, URL: https://www.iea.org/ [accessed: 2019-10-02].

[2] "Statistical office of the european union," European Commission, 2019, URL: https://ec.europa.eu/eurostat [accessed: 2019-10-02].

[3] K. C. Armel, A. Gupta, G. Shrimali, and A. Albert, "Is disaggregation the holy grail of energy efficiency? the case of electricity," Energy Policy, vol. 52, 2013, pp. 213–234.

[4] T. Alliance. Projet tbh. [Online]. Available: https://www.projet-tbh.fr/ (2015)

[5] L. G. Swan and V. I. Ugursal, "Modeling of end-use energy consumption in the residential sector: A review of modeling techniques," Renewable and Sustainable Energy Reviews, vol. 13, no. 8, 2009, pp. 1819–1835.

[6] K. Amarasinghe, D. L. Marino, and M. Manic, "Deep neural networks for energy load forecasting," in 2017 IEEE 26th International Symposium on Industrial Electronics (ISIE), June 2017, pp. 1483–1488.

[7] S. Seyedzadeh, F. P. Rahimian, I. Glesk, and M. Roper, "Machine learning for estimation of building energy consumption and performance: a review," Springer International Publishing, 2018.

[8] F. S. et al., "Predictive models for building's energy consumption: An artificial neural network (ann) approach," in 2015 XVIII AISEM Annual Conference, Feb 2015, pp. 1–4.

[9] P. A. Gonzlez and J. M. Zamarreo, "Prediction of hourly energy consumption in buildings based on a feedback artificial neural network," Energy and Buildings, vol. 37, no. 6, 2005, pp. 595–601.

[10] R. Platon, V. R. Dehkordi, and J. Martel, "Hourly prediction of a building's electricity consumption using case-based reasoning, artificial neural networks and principal component analysis," Energy and Buildings, vol. 92, 2015, pp. 10–18.

[11] J. Yuan, C. Farnham, C. Azuma, and K. Emura, "Predictive artificial neural network models to forecast the seasonal hourly electricity consumption for a university campus," Sustainable Cities and Society, vol. 42, 2018, pp. 82–92.

[12] A. S. Toms, M. Sumner, S. Lamparter, and D. Robinson, "On the energy demands of small appliances in homes," Energy Procedia, vol. 78, 2015, pp. 3384–3390, 6th International Building Physics Conference, IBPC 2015.

[13] I. Laicane, D. Blumberga, A. Blumberga, and M. Rosa, "Reducing household electricity consumption through demand side management: The role of home appliance scheduling and peak load reduction," Energy Procedia, vol. 72, 2015, pp. 222–229, international Scientific Conference Environmental and Climate Technologies, CONECT 2014.

[14] N. G. Dlamini and F. Cromieres, "Implementing peak load reduction algorithms for household electrical appliances," Energy Policy, vol. 44, 2012, pp. 280–290.

[15] R. Dhulst, W. Labeeuw, B. Beusen, S. Claessens, G. Deconinck, and K. Vanthournout, "Demand response flexibility and flexibility potential of residential smart appliances: Experiences from large pilot test in belgium," Applied Energy, vol. 155, 2015, pp. 79–90.

[16] E. Grover-Silva, E. Magliaro, and J. L. Conte, "Electric load profile non-intrusive analysis to guide energy efficient behavior for residential clients," no. 163. Bucharest, Romania: 2019 IEEE PES Innovative Smart Grid Technologies Europe, Oct. 2019, pp. 1–5.

[17] D. P. Kingma and J. Ba, "Adam: A method for stochastic optimization." ICLR 2015: International Conference on Learning Representations, jan 2017.

[18] Enertech, "Lighting appliance mesurement compaigne for 100 households in france (campagne de mesures de l'eclairage dans 100 logements en france)," Electricite de France et Agence de l'Environnement et de la Maitrise de l'Energie, Tech. Rep., 2004.

[19] T. Kreitz, "Refrigeration, washing machines and ac appliance mesurement compaigne (campagne de mesures des appareils de production de froid, des appareils de lavage et de la climatisation)," ADEME, Tech. Rep., 2016.

[20] Enertech, "Load mesurements for residential multi-media appliances and information technology devices (mesure de la consommation des usages domestiques de laudiovisuel et de linformatique)," ADEME, Tech. Rep., Jul 2008.

[21] O. Sidler, "Knowledge and management of electrical appliances for the residential sector (connaissance et maitrise des usages specifiques de l'electricite dans le secteur residentiel)," Enertech, Tech. Rep., 2009.

[22] R. Y. et al., "Quantifying flexibility of commercial and residential loads for demand response using setpoint changes," Applied Energy, vol. 177, 2016, pp. 149–164.

[23] D. Alexander, A. Tropsha, and D. A. Winkler, "Beware of R^2 : Simple, unambiguous assessment of the prediction accuracy of QSAR and QSPR models," Journal of Chemical Information and Modeling, vol. 55, 2015, pp. 1316–1322.

Predictive Model for Power Outages

Vivian Sultan
California State University
College of Business and Economics
Los Angeles, CA 90032
Email: vsultan3@calstatela.edu

***Abstract*—This study aims to address the question "How to predict power outages?" A statistical model in Statistical Package for Social Sciences (SPSS) is used to predict power-outage-event duration and customer calls using a stepwise regression algorithm. The model presented in this study can help advance smart-grid reliability by predicting power outages and taking the necessary steps to prevent them. Future work may involve enhancing the model's success and adding significant predictive variables.**

***Keywords*—data analytics; power-failure; smart-grid.**

I. INTRODUCTION

In a short time, electrical power has become a necessity of modern life. Our work, healthcare, leisure, economy, and livelihood depend on the constant supply of electrical power. Even a temporary power outage can lead to relative chaos, financial setbacks, and possible loss of life. U.S. cities dangle on electricity and, without a constant supply from the power grid, pandemonium would ensue. Power outages can be especially tragic when they endanger life-support systems in hospitals and nursing homes or systems in synchronization facilities such as airports, train stations, and traffic control. The economic cost of power interruptions to U.S. electricity consumers is $79 billion annually in damages and lost economic activity [1]. In 2017, Lawrence Berkeley National Laboratory provided an update, estimating power-interruption costs have increased more than 68% per year since their 2004 study [2].

Many reasons underlie current power failures. Among these reasons are severe weather, damage to electric transmission lines, shortage of circuits, and the aging of the power-grid infrastructure. Severe weather is the leading cause of power outages in the United States [3]. In 2019, weather events as a whole cost U.S. utilities $306 billion: the highest figure ever recorded by the federal government [4].

The aging of the grid infrastructure is another noteworthy reason for power failures. In 2008, the American Society of Civil Engineers gave the U.S. power-grid infrastructure an unsatisfactory grade [5]. They stated in a report that the power-transmission system in the United States required immediate attention. Furthermore, the report mentioned that the U.S. electric-power grid is similar to those of third-world countries. According to the Electric Power Research Institute (EPRI), equipment such as transformers controlling power transmission need to be replaced, as they have exceeded their expected lifespan considering the materials' original design [6].

Electrical outages have three main causes: (1) hardware and technical failures, (2) environment-related, and (3) human error [7]. Hardware and technical failures are due to equipment overload, short circuits, brownouts, and blackouts, to name a few [8]–[10]. These failures are often attributed to unmet peak usage, outdated equipment, and malfunctioning back-up power systems. Environment-related causes for power outages comprise weather, wildlife, and trees that come into contact with power lines. Lightning, high winds, and ice are common causes of weather-related power interruptions. Also, squirrels, snakes, and birds that come in contact with equipment such as transformers and fuses can cause equipment to momentarily fail or shut down completely [8]. As for the third main cause for electrical outages, human error, the Uptime Institute estimated that human error causes roughly 70% of the problems that plague data centers. Hacking can be included in the human-error category [11].

Analytics have been a popular topic in research and practice, particularly in the energy field. The use of analytics can help advance smart-grid reliability through, for example, elucidating a root cause of power failure, defining a solution for a blackout through data, or implementing a solution with continuous monitoring and management. In this research paper, the aim is to unveil the novel use of data analytics in predicting power-outage-event duration and customer calls. As the objective in this research is to advance smart-grid reliability, this paper explores ways to create a predictive model for power outages.

II. DATA SELECTION AND METHODOLOGY

EPRI's data repository includes the primary datasets used to conduct this analysis. The data sets include data from advanced metering systems, supervisory-control and data-acquisition systems, Geographic Information Systems, outage-management systems, distribution-management systems, asset-management systems, work-management systems, customer-information systems, and intelligent electronic-device databases. Access to datasets was provided as part of EPRI's data-mining initiative; the initiative provides a test bed for data exploration and innovation and seeks to solve major challenges faced by the utility industry [12].

To further enhance EPRI's dataset, data from other institutions are collected and aggregated. Specifically, Georgia Spatial Data Infrastructure and the Georgia GIS Clearinghouse are the sources for monthly temperature and precipitation data [13]. Additional data regarding storm events and storm details come from the National Oceanic and Atmospheric Administration website (NOAA) [14]. The data size is about 76,000 outages with 13 attributes.

The first step of the project methodology was to load data files from EPRI's Data Repository along with all aforementioned weather data to ArcGIS, the Geographic Information Systems platform created by ESRI. To streamline and make sure that the process is reusable and repeatable, ModelBuilder tool in ArcGIS is utilized to design data workflow. The workflow models spatially join the 48 map layers of weather data (from the Georgia Spatial Data Infrastructure and the Georgia GIS Clearinghouse website) with the outage map layer provided by EPRI. This serves as a final dataset for the study. Next, data exploration through correlational analysis in SPSS and GeoDa software has been conducted. The final step was to run a stepwise regression in SPSS.

Prior to all statistical analyses, data preparation follows these steps:

- Several variables need expert interpretation. For instance, the following variables: forestry expected pruning staff hours, average standard tree-pruning miles with bucket, average mechanical tree-pruning miles, average climbing tree-pruning miles, and actual pruning staff hours/circuit mile, had missing data. Aligned with the expert's instruction, missing data for these variables is replaced with a zero (0).
- The variable pole age had missing data. Per the instruction of the client's expert, transformer age replaces the missing data on pole age, with the reasoning that poles and transformers are routinely installed in tandem.
- The following variables: average climbing tree pruning miles, average standard tree-pruning miles with bucket, average mechanical tree-pruning miles, and forestry expected pruning staff hours either perfectly correlated (r = 1.00) with each other or nearly perfectly correlated ($r > 0.90$) with each other. Thus, these variables injected multicollinearity issues into the stepwise-regression equations. Due to the high level of multicollinearity, all of these variables are removed from the regression equation except one, forestry expected pruning staff hours.
- A stepwise regression algorithm is employed to create two regression equations. As Vogt [16] notes, researchers use a stepwise regression algorithm to find the "best" equation possible when regressing a dependent variable onto multiple independent variables. In other words, only sstatistically significant predictors of the dependent variable in a stepwise regression.

III. Analysis and Results

A. Descriptive Statistics in SPSS

Figures 1 and 2 were the outcomes of the initial data exploration of power-outage events. Percentages and frequencies were calculated for two main categories of outages: either based on storm event or due to forestry management. The breakdown is shown in Table I.

Ritchey [15] notes that for categorical variables, percentages and frequencies are the appropriate descriptive statistics to report. A statistical summary was calculated for all continuous variables in the sample. These data appear in Table II.

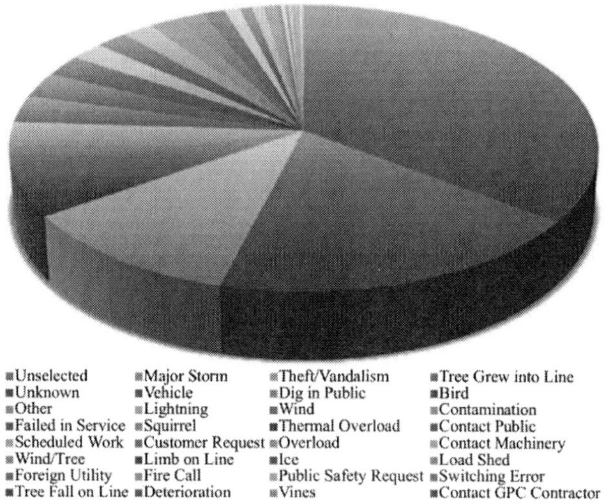

Figure 1. Reported outage events percent count by cause.

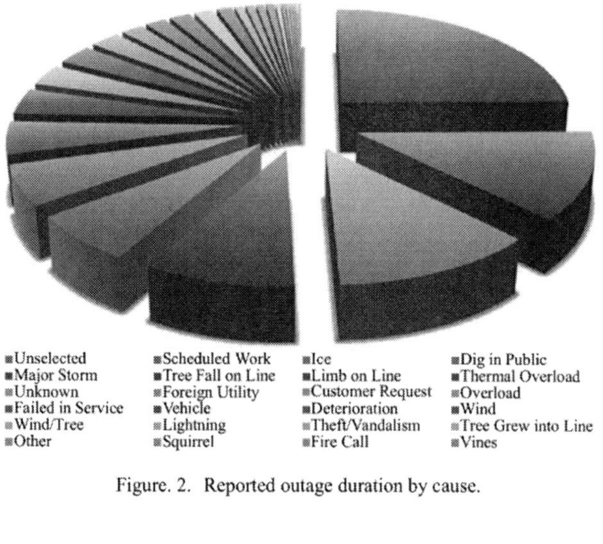

Figure. 2. Reported outage duration by cause.

TABLE I. PERCENTAGES AND FREQUENCIES, STUDY VARIABLES

	Frequency	Percent
Storm event		
Yes	59,078	76.9%
No	17,769	23.1%
Forestry management		
Yes	47,175	61.4%
No	29,672	38.6%
n	76,847	100.0%

Ritchey [15] notes that for continuous variables, means and standard deviations are the appropriate descriptive statistics to report.

B. Correlational Results

Results of correlation analysis in SPSS indicated a strong positive correlation between variables, such as storm events and outage-event duration . As expected, a storm event, precipitation, older poles, higher forestry expected pruning human hours, and higher levels of transformer age increased outage-event duration. A negative correlation between variables appeared between variables such as forestry management and outage-event customer calls. Surprisingly, engaging in forestry management, having older poles, and having lower levels of actual pruning human hours/circuit mile decreases the number of outage-event customer calls.

TABLE II. MEANS AND STANDARD DEVIATIONS, STUDY VARIABLES

Variable	M	SD	Min	Max
Outage event duration (minutes)	89.15	204.81	0	3589
Outage event customer calls	11.18	85.07	0	4888
Temperature (mean)	62.52	13.72	40.25	80.75
Precipitation	4.29	0.66	2.8	5.80
Forestry expected pruning man hours	858.01	882.24	0	3300
Average standard tree pruning miles with bucket	6.63	6.48	0	20.49
Average mechanical tree pruning miles	3.00	2.94	0	9.28
Average climbing tree pruning miles	0.75	0.73	0	2.32
Actual pruning man hours / circuit mile	42.18	36.80	0	157
Transformer age	4.50	1.86	3	8
Pole age	23.90	16.76	3	93

Note: n = 76,847.

Statistically significant correlations were flagged in the correlation table (Table III) in the following manner:

- A single star (*) denotes a significant correlation at the $p = .05$ alpha level.
- A double star (**) denotes a significant correlation at the $p = .01$ alpha level.
- No stars means the correlation is not statistically significant, and no relationship exists among the two variables in question.
- An inverse correlation was denoted by a negative sign (-). An inverse correlation means that as one variable increases in value, the other variable decreases in value.
- A positive correlation was denoted by the absence of a negative sign (-). A positive correlation means that as one variable increases in value, the other variable increases in value.

TABLE III. CORRELATION RESULTS

	1	2	3	4	5	6	7	8	9	10	11	12
2	0.09**	1.00										
3	0.08**	0.01	1.00									
4	-0.13**	0.01	0.14**	1.00								
5	0.08**	0.00	-0.06**	-0.37**	1.00							
6	0.01	-0.02**	-0.01	0.00	-0.03**	1.00						
7	0.01**	0.00	0.00	0.00	-0.02**	0.77**	1.00					
8	0.01**	0.00	0.00	0.00	-0.02**	0.81**	0.96**	1.00				
9	0.01**	0.00	0.00	0.00	-0.02**	0.81**	0.96**	1.00	1.00			
10	0.01**	0.00	0.00	0.00	-0.02**	0.81**	0.96**	1.00**	1.00	1.00		
11	0.01*	-0.01**	-0.01	-0.01**	-0.02**	0.91**	0.85**	0.79**	0.79**	0.79**	1.00	
12	0.01*	0.01**	0.01**	0.00	0.02**	-0.13**	-0.11**	-0.12**	-0.12**	-0.12**	-0.11**	1.00
13	0.02**	-0.01**	0.02**	-0.01**	0.03**	0.02**	0.04**	0.03**	0.03**	0.03**	0.03**	-0.03**

Note: 1. Outage-event duration; 2. Outage-event customer calls; 3. Storm event (1 = *yes*); 4. Temperature (mean); 5. Precipitation; 6. Forestry management (1 = *yes*); 7. Forestry expected pruning human hours; 8. Average standard tree-pruning miles with bucket; 9. Average mechanical tree-pruning miles; 10. Average climbing-tree-pruning miles; 11. Actual pruning human hours/circuit mile; 12. Transformer age; 13. Pole age; $^{*}p < .05$; $^{**}p < .01$, two-tailed tests.

C. Multiple Linear Regression Results

As Ritchey [15] notes, a multiple linear regression technique is appropriate when the dependent variable is continuous in nature and two or more independent variables are in use. The current circumstances satisfy these criteria. The idea of stepwise regression is to add all independent variables into a regression equation that relates to the dependent variables. Then, the process involves iteratively peruse the regression, removes the variables that are not statistically contribute to the dependent variable. In this paper, we have 2 dependent variables of interest: outage-event customer calls and outage-event duration. These dependent variables will result in two regression equations that will be described below.

Table IV presents the results of the stepwise multiple linear regression of outage-event customer calls onto the several independent predictors. The Omnibus F-test, shown in Table IV, is statistically significant ($F = 18.217$; $df = 5, 76841$; $p < .001$). Thus, the decomposition of effects in the regression model can proceed.

TABLE IV. MULTIPLE LINEAR REGRESSION OF OUTAGE EVENT CUSTOMER CALLS ONTO THE PREDICTORS

Variable	*B*	*SE(B)*	*p*
Constant	12.214	1.044	0.000
Forestry management	-4.084	1.515	0.007
Forestry expected pruning man hours	0.004	0.001	0.000
Pole age	-0.072	0.018	0.000
Transformer age	0.499	0.167	0.003
Actual pruning staff hours/circuit mile	-0.068	0.024	0.004
N	76847		
F	18.217		0.000
Adjusted R^2	0.001		

Based on the significance of the table, five variables have been retained using the stepwise regression algorithm. Three of these variables lower the number of outage-event customer calls. Specifically, engaging in forestry management ($B = -4.084$, $p = .007$), having older polls ($B = -0.072$, $p < .001$), and having lower levels of actual pruning staff hours/circuit mile ($B = -0.068$, $p = .004$) decrease the number of outage-event customer calls. Two of the variables raise the number of outage-event customer calls. Specifically, having higher levels of forestry expected pruning staff hours ($B = 0.004$, $p < .001$) and higher levels of transformer age ($B = 0.499$, $p = .003$) increase the number of outage-event customer calls. The adjusted R^2 value was identical to the R^2 value.

Table V presents the results of the stepwise multiple linear regression of outage-event duration onto the several independent predictors. The Omnibus F-Test in Table V is statistically significant ($F = 218.672$; $df = 5, 76841$; $p < .001$). Thus, decomposition of effects in the regression model can proceed.

TABLE V. MULTIPLE LINEAR REGRESSION OF OUTAGE EVENT DURATION ONTO THE PREDICTORS

Variable	B	SE(B)	p
Constant	-62.703	5.511	0.000
Storm event	42.413	1.744	0.000
Precipitation	24.796	1.112	0.000
Pole age	0.243	0.044	0.000
Forestry expected pruning man hours	0.004	0.001	0.000
Transformer age	0.898	0.398	0.024
N	76847		
F	218.672		0.000
Adjusted R^2	0.014		

A count of five variables were retained by the stepwise regression algorithm. All five variables raise the outage-event duration. Specifically, having a storm event ($B = 42.413$, $p < .001$), having precipitation ($B = 24.796$, $p < .001$), having older poles ($B = 0.243$, $p < .001$), having higher forestry expected pruning staff hours ($B = 0.004$, $p < .001$), and higher levels of transformer age ($B = 0.898$, $p = .024$) increase the outage-event duration.

IV. CONCLUSION

This study aimed to address "How to predict power outages." To address the research, A predictive novel model was developed in SPSS to predict the power-outage-event duration and customer calls. A stepwise regression algorithm was used for the two regression equations. The SPSS model presented in this study can help advance smart-grid reliability by predicting power outages and taking the necessary steps to prevent them. Future work may involve enhancing the model's success and adding significant predictive variables. Data analytics can be a major resource of assistance for managing power-failure events.

One limitation of this research is that pole-age data was used as a proxy for infrastructure age and the rest of the equipment data. Future work may involve connecting to virtually any type of streaming data feed and transforming data-analytics applications into frontline decision applications, predicting and updating power-outage incidents as they occur.

From this research, it was concluded that SPSS and GIS tools offers a solution to analyze the electric-grid distribution system. This model provides evidence that SPSS can perform the analysis to predict power-outage-event duration and customer calls. If additional funds and data are made available, one can expand on this analysis to create a custom solution for the utility industry to control and forecast power outages. Data analytics can be a major resource of assistance for electronic-inspection systems, to lower the duration of customer outages, to improve crew-response time, and to reduce labor and overtime costs.

REFERENCES

[1] K. LaCommare and J. Eto, "Understanding the cost of power interruptions to U.S. electricity consumers," https://emp.lbl.gov /sites/all/files/lbnl-55718.pdf

[2] J. Eto, "The national cost of power interruptions to electricity consumers —Revised update," http://grouper.ieee.org/groups/td /dist/sd/doc/2017-01-10 National Cost of Power Interruptions to Electricity Customers - Eto.pdf

[3] President's Council of Economic Advisers and the U.S. Department of Energy's Office of Electricity Delivery and Energy Reliability, "Economic benefits of increasing electric grid resilience to weather outages," http://energy.gov/sites/prod/files/2013/08/f2/Grid Resiliency Report_FINAL.pdf

[4] J. Porter, "The $306 billion question: How to make outage management better?" June 12, 2018, https://www.elp.com/Electric-Light-Power-Newsletter/articles/2018/06/the-306-billion-question -how-to-make-outage-management-better.html

[5] American Society of Civil Engineers, "Infrastructure fact sheet," 2009 http://www.infrastructurereportcard.org/2009/sites/default/files /RC2009_rail.pdf

[6] D. Stone, "It's the electric grid, stupid," September 9, 2011, http://www.thedailybeast.com/articles/2011/09/09/major-power-outage -shows-weakness-of-aging-electric-infrastructure.html

[7] K. Chayanam, "Analysis of telecommunications power outages—Inline," https://etd.ohiolink.edu/rws_etd/document/get/ohiou1125024491 /inline

[8] Westar Energy, "What causes power outages? – working to improve service reliability," https://www.westarenergy.com/outage-causes

[9] Rocky Mountain Power, "Key causes of power outages." https://www .rockymountainpower.net/ed/po/or/kcopo.html

[10] Diesel Service and Supply, "Causes of power failures & power outages," http://www.dieselserviceandsupply.com/Causes_of_Power _Failures.aspx

[11] R. Miller, "How to prevent downtime due to human error," https:// www.datacenterknowledge.com/archives/2010/08/13/how-to-prevent -downtime-due-to-human-error

[12] Electric Power Research Institute, "EPRI distribution modernization demonstration (DMD) data mining initiative," http://smartgrid.epri .com/DMD-DMI.aspx

[13] Georgia, G. I. S. Data Clearinghouse,"About the Clearinghouse." Online at www. gis. state. ga. us/Clearinghouse/clearinghouse. html.

[14] National Oceanic and Atmospheric Administration, Storm events database.", 2018, Online at https://www.ncdc.noaa.gov/stormevents/

[15] F. Ritchey, The statistical imagination: Elementary statistics for the social sciences, 2nd edition. Boston, MA: McGraw-Hill, 2008.

[16] W. P. D Vogt, Dictionary of statistics & methodology: A nontechnical guide for the social sciences, 2nd edition. Thousand Oaks, CA: Sage, 1999.

A Framework for Strategy Selection of Atomic Entities in the Holonic Smart Grid

Alexander Wallis, Sascha Hauke
University of Applied Sciences Landshut
Landshut, Germany
e-mails: {alexander.wallis, sascha.hauke}@haw-landshut.de

Rolf Egert, Max Mühlhäuser
Technical University of Darmstadt
Darmstadt, Germany
e-mails: {egert, max}@tk.tu-darmstadt.de

Abstract—**Maintaining the continuous supply of electricity within smart microgrids is a challenging task, which becomes increasingly difficult with the growing integration of volatile Renewable Energy Sources (RES). Quick changes within the production behavior of these resources can disturb the necessary balance between demand and supply and may ultimately lead to blackouts within the grid. To prevent balance disturbances, electricity production and consumption needs to be coordinated and power needs to be shared among the participants within the microgrid. Facilitating coordinated behavior of grid entities and ensuring reliable operation of the microgrid in the presence of volatile RES requires sophisticated strategies for operating individual participants. In this paper, we present a modular framework to support dynamic energy distribution for atomic entities (producers/consumers) in holarchically organized energy grids. In particular, the framework provides production and consumption forecasting to enable intelligent strategy selection to improve the day-ahead control decisions for atomic entities. The proposed framework enables the bottom-up formation of smart mircogrid holons and represents a foundation for the formation and strategic coordination of participants in smart microgrids.**

Keywords—Micro Grids; Holonic Smart Grids; Optimization; Forecasting; Strategy Selection.

I. INTRODUCTION

Electrical grids are evolving from a centrally managed critical infrastructure to distributedly managed Smart Grids (SGs). This evolution is driven by the need for the grid to incorporate local production capabilities of renewable Distributed Energy Resourcess (DERs). The paradigm shift from centralized to distributed control, however, leads to a considerable increase in the complexity of network management tasks. Various approaches for tight monitoring and fast control have been put forward to support continuous operation and provide stability of distributed energy grids [1]–[3]. These approaches generally rely on strong support by Information and Communication Technologies (ICT).

Hierarchical and cellular network segmentation promises to simplify the mechanisms for controlling the SG. The next evolutionary step for cellular network approaches are *holar* structures [4]. In particular, these systems seek to leverage formation and segmentation by enabling the reuse of mechanisms on different hierarchical levels. Entities in such a system (so-called Holons) are simultaneously a "whole" and a "part" of something bigger. The emerging system-of-systems structure is referred to as a holarchy [5]. Holons are dynamic cells, which can merge with other holons (or separate into individual smaller ones) when suitable. Under optimal conditions, holons tend to form larger holons, while their capability to separate sub-parts aids in increasing network stability (e.g., by splitting off potential misbehaving or faulty entities). This is ensured cause holocharies are mainly based on the concepts of *isolation* and *containment* [6].

In this work, we consider single buildings, be they commercial or residential, to be the atomic building blocks of holons. With the integration of DERs, they may be both producers and consumers, so called *prosumers* of energy. In order to facilitate holon creation and stable operation, particularly in small-scale grid scenarios, accurate models for the behavior of these prosumers are necessary. This, in turn, entails the need for a framework that is capable of forecasting, within reasonable limits, both electrical load and production behaviors. This need is exacerbated especially in small holons, i.e., smart microgrids, because smoothing effects on energy production and consumption are not as effective here as in larger grids.

Based on the considerations mentioned above, we present a framework for stable holon operation. In particular, at an atomic prosumer level, the framework consists of a consumer, production unit, storage system and a power supply. The main contributions of the proposed framework are:

- Provision of dynamic control via smart strategy selection for holonic smart microgrids.
- Advancement of current smart microgrid capabilities by enabling forecasting and operation optimization on the level of atomic holons.
- Showcasing the applicability of the current deployment of the framework by deploying it at a real-world prosumer site.

The rest of the paper is structured as follows: Section II describes the topology and power flow of an atomic holon, as well as the conceptual framework model and information transmissions. In Section IV, further details concerning the forecast model and operation strategy optimization are provided. We finalize this work with conclusion and future works in Section V.

II. System

The goal of this section is to present the system description of an atomic building block of holons and the energy flow between the different components. Following this, the structure and information flow of the framework is detailed.

A. Topology

To manage the merging and splitting process of a holon, an optimized energy flow at the atomic level, i.e., for individual consuming and producing participants, is necessary. Therefore, the individual components that are encompassed in an atomic holon require a clear definition. Figure 1 shows the four components of what we treat as an atomic holon: *consumer*, *producer* (PV), *storage* (batteries) and *power grid*. An atomic holon can, but does not need to, implement all components. The arrows indicate the direction of possible energy flow.

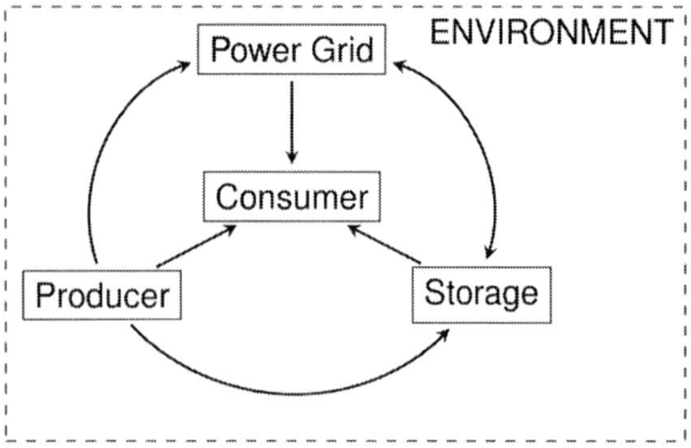

Figure 1. Energy flow between the four different holon components.

a) Power grid: The grid-connected power supply of the building. Ideally, the household's load is covered by batteries and direct consumption of self-produced energy (e.g., solar panel). In case of an increased demand, which cannot be compensated by the locally available resources, the remaining difference is taken from the general power grid.

b) Consumer: The total aggregated energy consumption of an atomic holon (This includes potential incident consumption that results from the battery storage, the solar panel and the grid connection).

c) Producer: As mentioned in Section I, every entity in the grid, which is capable of supplying electricity to itself or others is considered as an energy producer. As holons are envisioned to represent prosumers in the future energy grid the *producer* component represents the aggregated production capacity of the holon. For instance, single households with solar Photovoltaic (PV) cells or commercial buildings with wind turbines.

d) Storage: A main enabler for the efficient use of Renewable Energy Sourcess (RESs) is their combined used with an energy storage system. Aside from storing excess energy during times of high production (e.g., sunny days at noon), energy storage solutions are used to compensate for the volatile behavior of RESs. The resulting inconsistency of production behavior of RESs – due to intermittency of weather conditions – or daily consumption variations can be mitigated using energy storage systems. In general, these storage systems can consist of any type of batteries, including the batteries of attached electric vehicles.

B. Framework

The proposed framework aims to optimize control strategies in systems that are structured according to the *holarchy* concept. In particular, we focus on system control of atomic holons within a holarchy. For these holons, the following characterizations are necessary [6]:

- *Autonomy*: As holons are, simultaneously a *part* and a *whole* within a system and they may have individual goals that may differ from the general goals of the system as a whole. A holon's *autonomy* property describes its striving to fulfill its own objectives. For this, it is capable of making decisions and to create and manage the execution of its own plans and strategies.
- *Self-contained*: Each holon is a *whole* itself and can exist or work without external input by exploiting its own resources.
- *Co-operation*: Holons can cooperate on the basis of special communication and interaction rules. Holons can cooperate with other holons on similar levels within a holarchy. Additionally, they can merge with other holons to form larger holons or split into smaller holons. This concept of cooperation enables holons to achieve more complex goals (e.g., due to more available resources after a merging process), and enables to interrupt the cooperation by isolating faulty parts.

In order to facilitate holarchical operation in a better way, the proposed framework uses a bottom-up approach. Instead of organizing holons with a top-down approach, this framework regulates the strategic operation within an atomic building block, composed of the previously described system's primary components. In future work, we will expand the prediction and strategy selection to non-atomic holons.

The framework architecture can be seen in Figure 2. It is divided into three main parts: *Forecasting of Resources for Dynamic Optimization (FRODO)*, *Optimal Load and Energy Flow (OLAF)*, and *Environment*. As depicted in the architecture layout, the framework works based on historical data of power consumption as well as production. Dashed-line arrows indicate the flow of information between the framework components. FRODO receives historical data and derives two forecasts: PV production and load consumption. Information about these two forecasts, in addition to information about the holar system structure, are then used by OLAF as an input for the *strategy selection process*. This selected strategy is the foundation for next days charging and discharging schedule of a storage unit within an atomic holon. After considering the related work in the following Section III, Section IV provides detailed information for the individual parts of the framework.

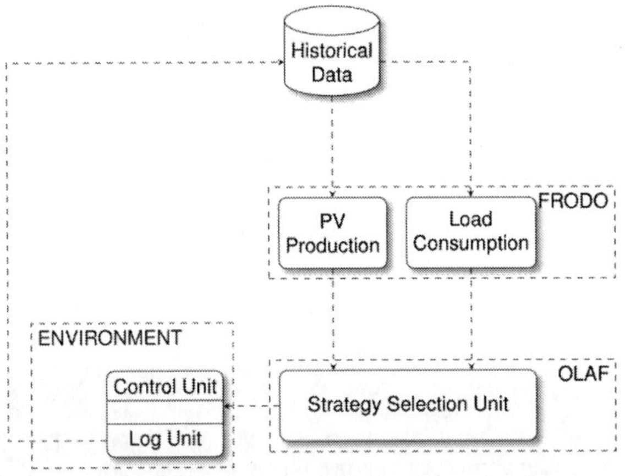

Figure 2. Structure and information flow within the Framework.

III. Related Work

In related work, Battery Energy Storage Systems (BESSs) for handling fluctuations in DER are a major research topic. The requirements of a BESS for mitigating PV output fluctuations are examined in [7]–[9]. In [10], a control scheme for energy devices in the distribution network to reduce peak demand based on day-ahead demand forecasts is presented. Thereby, the research focuses on optimizing the storage system's schedule from the network operator's perspective. [11] proposes a solution for finding the optimal size of a BESS with regard to economic perspective, which can be used in addition to finding the optimal charge and discharge strategy under certain conditions. In [12], a day-ahead energy management framework of a microgrid is presented; the authors propose management of the energy flow based on next days electricity price. That is one possible strategy for controlling the energy flow. In this work, we are defining several strategies and looking for the best based on external constraints. Batteries are also used in [13] to maximize a households profit. Control strategies are applied for scheduling electric vehicles to implement peak-shaving and valley-filling in [14]. In this scenario, the electric vehicle takes part of a storage unit in Figure 1.

IV. Methods

The promising results in BESS-oriented related work suggest the potential of prediction-based strategy selection for the dynamic creation of hólar microgrids. The framework proposed in this paper aims to leverage these results and, in future work, also to integrate formation and control methods for merging and splitting holar microgrids.

This section gives detailed information for the three different parts of the framework: *Historical Data* (IV-A), *FRODO* (IV-B) and *OLAF* (IV-C). Additionally, we show the first results on day-ahead load forecasting based on real-world data.

A. Historical Data

The load consumption and PV production data used in this paper are provided by the *Technology Centre of Energy (TZE)* at the Landshut University of Applied Sciences. The load consumption and PV production data was collected from January 1st, 2017 to December 31st, 2018 and are available for every minute. However, in this paper we are using 60 minutes – or hourly – discrete power values for production [kW] as well as consumption [kW]. Weather data are recorded at the stationary installed weather station at the TZE for *temperature* [°C], *humidity* [%], *solar irradiation* [W/m^2], *precipitate* [mm/min], *wind speed* [m/s], *wind direction* [°] and *air pressure* [hPa]. Missing weather data points are taken from the *Deutscher Wetterdienst (DWD)*.

To ensure valid forecasting results by FRODO, the following data preprocessing steps are implemented:

- Data cleaning: To reduce the influence of missing data, interpolation was executed if the gap is less than one hour. Otherwise the average value for the specific time-slot was scaled to fit the curve. Since we are doing day-ahead forecasting, days without previous load consumption are also removed.
- Time-series to supervised learning: The raw data is a time-series sequence ordered by a time index for every minute, which is first aggregated to hourly discrete values and then converted to input-output-pairs (x_n, t_n) where x_n are the inputs and t_n the output values for each day $n = 1...N$.
- Input selection: Feature engineering is a crucial task in machine learning. Figure 3 shows the daily average power consumption per weekday and there is a significant difference between workday and weekend visible. Therefore, the date features are *one-hot-encoded*, i.e. 1000000 for Monday, 0100000 for Tuesday and so on. The same applies to special days like *holidays*.

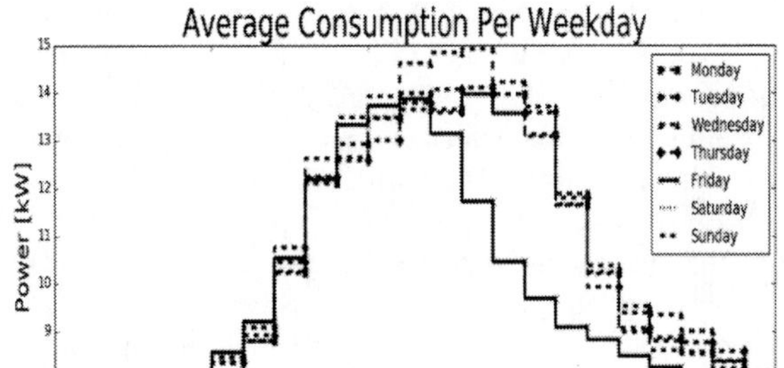

Figure 3. Daily average consumption per weekday over a 2-year period.

For illustration, Figure 3 depicts the daily discrete power values for consumption over a day divided into 23 hourly intervals. Beside the previous mentioned difference between workday and weekend, there is a clear decrease after Friday lunchtime visible. This is explained due to the fact that the *consumer* in this show-case is formed by a research center with around 15-20 employees working regularly on weekdays. The *production* unit in this research is a 10.4 kWp installed PV module and different batteries, e.g., 100 kWh Redox-Flow, 125 kWh Lithium-Ion, 12 kWh Lithium-Ion, are used as *storage* systems. These are the key components for the atomic holon in this show-case as well as the consumption data – and also the production data – which provides the basis for the forecasting module, described in the following part.

B. FRODO

To ensure an efficient energy flow between the different entities within a holon, a precise load consumption and PV production forecast is essential. Due to page limitations, only the development process for the load consumption component within FRODO is described. The PV production forecasting works similarly and will be shown in subsequent work. Different forecast approaches can be classified according to their forecast horizon. Hereby, we differentiate between three main categories: *Long-Term Load Forecast* (1 year to 10 years ahead), *Medium-Term Load Forecast* (1 month to 1 year ahead), *Short-Term Load Forecast* (1 hour to 1 day or 1 week ahead) [15]. A main requirement for using the proposed framework is to be able to dynamically select the next day's operation strategy. For enabling this, a Short-Term Load Forecast (STLF) is essential. Therefore, two different Machine Learning (ML) models for STLF are developed and tested against each other: A first model leverages *Random Forest (RF)* and the second one is based on a *Long-Short-Term memory (LSTM) neural network*. Both approaches have been tested and proven to work well for the present forecasting task in the related literature [16]–[19].

Architecture: The RF is an ensemble method that operates by constructing a multitude of decision trees whereby each tree forecasts the load consumption by itself and the method returns the mean value. In this case, 500 trees are created at training time with a maximum depth of 15. The LSTM is a recurrent neural network and unlike standard feedforward networks, LSTM has feedback connections to maintain information, which is used for solving learning tasks based on prior input data and decisions [20]. We use one hidden layer with 50 neurons, one dropout layer with 0.2 rate to avoid over-fitting and one dense layer with 24 neurons, since we want forecast values for every hourly interval. The prior mentioned hyperparameters of each method are estimated through extensive *grid search*. For that reason, the historical data are first split into three distinct groups: *training* (80 %), *validation* (10 %) and *test* (10 %). The first two are used for tuning the previously mentioned hyperparameters. After estimating the best model architecture, each model is tested by using a *10-Fold Cross-Validation*.

Input Features: In [21], we showed that the power consumption for an exemplary day d can be explained by the consumption values of the previous day $d-1$ and by last week's consumption $d-7$. In addition, we mentioned previously the importance of calendric factor like day of week and holidays. Therefore, the input features are composed of the following 58 variables: (*Input* $1-24$) consumption of $d-1$ [kW], (25) mean temperature T_{d-1} [°C], $(26-50)$ consumption of $d-7$ [kW], (51) mean temperature T_{d-7} [°C], $(52-57)$ one-hot-encoded day of week and (58) holiday [0 or 1]. Training, validation and test data are input-output pairs (x_n, t_n), with $n = 1...N$, where x_n are the explanatory input variables and t_n the outputs.

Evaluation: The test data is used to evaluate the model's forecast accuracy. Therefore, the third split of the historical data was held back to measure the model's accuracy on "unknown" input data. Figure 4 shows next day's consumption forecast of both RF and LSTM for one example day, and also the actual measured values for comparison. It shows good accuracy in the morning as well as in the evening, but a clear underestimation for the three intervals between 11 and 14 o'clock. This behavior is caused by the fact that those consumption values are way above average (see Figure 3) and can not be handled well by both RF and LSTM. To improve this forecast behavior, further model adjustments are done in future work described in Section V.

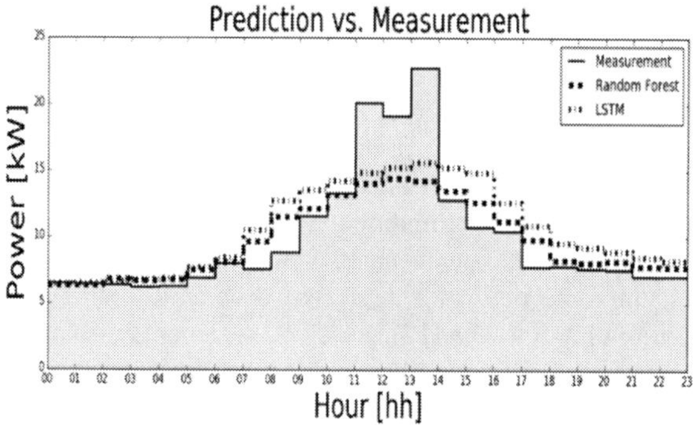

Figure 4. Forecast values and actual consumption for one example day.

For model comparison, three different persistence approaches are used. Let $L(d, h)$ be the consumed load for a specific day d at a particular hour h. A persistence model is a method to predict the future behavior on the assumption that current load consumption $L(d_t, h)$ is similar to the value at the same hour of a different day $L(d_{t-i}, h)$. The following approaches are implemented: (1) *"previous day"* $L(d_t, h) = L(d_{t-1}, h)$, (2) *"last week"* $L(d_t, h) = L(d_{t-7}, h)$ and (3) *"weekday average"* $L(d_t, h) = \overline{L(d_t, h)}$, where $\overline{L}$ denotes the average value for a specific day and hour.

To quantify the results numerically, forecast accuracy is presented using standard measures: Mean Squared Error (MSE), Root Mean Squared Error (RMSE), Mean Absolute Error (MAE) and Mean Percentage Error (MPE), defined in the following equations Eq. (1)–(4):

$$MSE = \frac{1}{N}\sum_{n=1}^{N}(Y_n - \widehat{Y_n})^2 \quad (1)$$

$$RMSE = \sqrt{MSE} \quad (2)$$

$$MAE = \frac{1}{N}\sum_{n=1}^{N}\left|Y_n - \widehat{Y_n}\right| \quad (3)$$

$$MPE = \frac{100\,\%}{N}\sum_{n=1}^{N}\left(\frac{Y_n - \widehat{Y_n}}{Y_n}\right), \quad (4)$$

where Y_n represents the actual value of the electrical load and parameter $\widehat{Y_n}$ is the corresponding forecast value, respectively. N is the sample size. The forecast results on the test set are given in Table I.

TABLE I. RESULTS OF MEASUREMENTS FOR DIFFERENT CONDUCTED FORECASTS

Model	MSE	RMSE	MAE	MPE
Previous Day	15.82	3.98	2.46	-6.51
Last Week	9.27	3.04	2.16	*1.45*
Weekday Average	5.42	2.33	1.77	-3.39
RF	*4.86*	*2.20*	*1.43*	-1.57
LSTM	6.23	2.45	1.67	-4.29

The results in Table I show that the RF has an increased forecast accuracy compared to the persistence approaches and that the LSTM performs worse than the "weekday average" persistence models. For every metric, the lowest value is the best result and is indicated by emphasized cells. It is also worth mentioning, that the forecast error for the RF remains mostly constant regardless of the different splits of the data set. In contrast, the LSTM results depend highly on the segmentation of training and test sets. This might be due to the LSTM overfitting of the data based on the distribution of workdays, weekends and outliers (way above average days). Due to page limitations, this section only describes the load consumption forecast. However, the production forecast (e.g., solar, wind) works similarly and will be the focus of further work. The derived energy production and consumption forecasts are further processed by OLAF, where the optimal strategy for the energy flow between the different holon components is selected. This selection process and also definitions of example strategies are explained in the following section.

C. OLAF

The OLAF module of the framework is responsible for selecting and executing the next day's operation strategy. Basically, this strategy is a sequence of information how and in which direction the energy flow between the components (see Figure 1) is realized for each next day's hourly interval. This includes schedules for charging and discharging the storage unit, based on some prior defined constraint and optimization goals, which are specified in the following paragraph. For this, OLAF receives the two previously generated PV production and load consumption forecasts of the FRODO module and selects a BESS operation strategy based on this information. The different possible BESS strategies can roughly be divided into three categories: *customer-*, *market-* and *grid-oriented* strategies [22]. A more detailed classification for operation strategies is presented in [23]. These strategies and the primary beneficiary are defined as follows:

- *Maximized consumption of self-generated power* (customer): Produced power is primary used to cover the household's load. Overproduction is going to the storage unit.
- *Limited power grid feed-in* (grid): Cuts off power feed-in at a given upper boundary.
- *Time-scheduled (dis-)charging* (grid/customer): The storage, e.g., Battery, is only used at certain time of days.
- *Time and power constrained storage* (grid): Minimize feed-in at given peak-hours for grid relief through empirical knowledge.
- *(Dis-)charging depending on energy pricing* (market/customer): This strategy's goal is to minimize the electricity price for a given household.
- *Incremental grid relief* (grid): Instead of feed-in as much produced power as possible, only a fixed percentage is used, e.g., 30 %.
- *State-of-Charge dependent charging* (customer): The charging schedule is prepared, depending on the battery's State Of Charge (SOC). The lower the SOC, the more produced power is used to charge the battery.

The *Strategy Selection Unit* encompassed within OLAF chooses, based on the forecast values and the desired holarchical organization, a preferred operation strategy. Since we are getting next day's forecast as hourly intervals from FRODO, the chosen strategy is executed also in hourly time steps. Subsequently, the *Control Unit (CU)* – an independent controlling element implemented in the environment – is responsible for executing the a-priori defined charge and discharge schedules for the respective storage systems (e.g., batteries) within each time interval. If the available energy does not satisfy the chosen strategy, e.g., discharge an empty battery or feed-in produced power although the limit is reached, the *Control Unit* attempts to adjust the next time interval – using the feedback loop – so that the desired goal can be achieved.

Although the *Log Unit (LU)* as well as the *CU* are not explicitly part of OLAF, they are both in communication with each other at the end/beginning of each time interval. One key function of the CU is to give feedback to OLAF, not only for strategy monitoring reasons, but also to adjust the forecast results from FRODO. If the forecast value for a time slot exceeds some threshold – both directions: overestimation or underestimation – the CU informs OLAF to modify the remaining daily values. Furthermore, the LU is responsible for logging and storing the energy data, the recorded weather data (either through a stationary station or external sources) and the operational data on a daily basis. This information is then stored as historical data and can then used by the FRODO module to improve the forecast accuracy for subsequent days, if the desired accuracy is not fulfilled anymore. This possible improvement – either update the forecast values or change the actual operation strategy – represents an ability to dynamically handle uncertain behavior, as it is necessary for the safe and resilient operation of energy grids base on the concept of holarchies and for performing the characterization described in Section II-B. Through smart strategy selection, e.g., *self-containment* for one atomic holon or *co-operation* between multiple atomic entities is achievable.

V. CONCLUSION

In this paper, we introduced a framework to provide dynamic control for the state-of-the-art holonic smart grid. Based on a bottom-up approach, the proposed framework enables holarchical organization at an atomic level. The presented approach is designed to improve current Smart Grid capabilities by providing a modular structure for forecasting and optimization. Based on historical load data, our framework makes a day-ahead forecast for load consumption and PV production, within reasonable limits. After analyzing the historical data (Section IV-A) an LSTM approach, which is an established ML technique, is presented in Section IV-B as the STLF model within FRODO. Due to page limitations, the PV production forecast was not presented in this research paper, but works similarly to the load consumption forecast and will be described in further works. To validate the load forecast accuracy, an RF and an ML technique are evaluated and compared, as well as the following three persistence models: previous day, last week and average weekday value. The results showed that both LSTM and RF are practicable methods with higher accuracy than the persistence approaches. The derived forecast values are the basis for OLAF, the optimization part within the framework, which selects the strategy for next day's energy flow within a holon. This operation strategy is chosen under predefined constraints, e.g., customer-oriented, market-oriented, grid-oriented, and is executed in one-hour-intervals over the next day. The framework is capable of handling uncertain behavior and divergent forecasts through a feedback-loop after each interval.

In future research work, we are improving the forecast models by adjusting the different ML techniques. Furthermore, we are defining more strategies to optimize the holarchical operation.

ACKNOWLEDGMENT

The authors would like to thank Martina Hörmann and Konstantin Ziegler for suggesting the naming for FRODO and OLAF. We also thank the TZE for providing the relevant data.

REFERENCES

[1] P. D. Curtis and N. Mehravari, "Evaluating and improving cybersecurity capabilities of the energy critical infrastructure," in *2015 IEEE International Symposium on Technologies for Homeland Security (HST)*. IEEE, Apr 2015, pp. 1–6.

[2] S. Kakran and S. Chanana, "Smart operations of smart grids integrated with distributed generation: A review," *Renewable and Sustainable Energy Reviews*, vol. 81, pp. 524–535, Jan 2018.

[3] E. Mengelkamp *et al.*, "Designing microgrid energy markets," *Applied Energy*, vol. 210, pp. 870–880, Jan 2018.

[4] M. Calabrese, A. Amato, V. di Lecce, and V. Piuri, "Hierarchical-granularity holonic modelling," *Journal of Ambient Intelligence and Humanized Computing*, vol. 1, no. 3, pp. 199–209, 2010.

[5] R. Egert, C. G. Cordero, A. Tundis, and M. Mühlhäuser, "HOLEG: A simulator for evaluating resilient energy networks based on the Holon analogy," in *Proceedings - 2017 IEEE/ACM 21st International Symposium on Distributed Simulation and Real Time Applications7*. IEEE, Oct 2017, pp. 1–8.

[6] H. Van Brussel, J. Wyns, P. Valckenaers, L. Bongaerts, and P. Peeters, "Reference architecture for holonic manufacturing systems: PROSA," *Computers in Industry*, vol. 37, no. 3, pp. 255–274, 1998.

[7] A. A. Jahromi *et al.*, "Battery Energy Storage Requirements for Mitigating PV Output Fluctuations," *Proceedings - 2018 IEEE PES Innovative Smart Grid Technologies Conference Europe, ISGT 2018*, pp. 1–5, 2018.

[8] J. Duan *et al.*, "Reinforcement-Learning-Based Optimal Control of Hybrid Energy Storage Systems in Hybrid AC–DC Microgrids," *IEEE Transactions on Industrial Informatics*, vol. 15, no. 9, pp. 5355–5364, 2019.

[9] J. Duan, H. Xu, W. Liu, J. C. Peng, and H. Jiang, "Zero-sum game based cooperative control for onboard pulsed power load accommodation," *IEEE Transactions on Industrial Informatics*, vol. 16, no. 1, pp. 238–247, 2020.

[10] M. Rowe *et al.*, "A Peak Reduction Scheduling Algorithm for Storage Devices on the Low Voltage Network," *Ieee Transactions on Smart Grid*, vol. 5, no. 4, pp. 2115–2124, 2014.

[11] R. Garmabdari, M. Moghimi, F. Yang, J. Lu, H. Li, and Z. Yang, "Optimisation of battery energy storage capacity for a grid-tied renewable microgrid," in *2017 IEEE Innovative Smart Grid Technologies - Asia (ISGT-Asia)*. IEEE, Dec 2017, pp. 1–6.

[12] D. Prudhviraj, P. B. Kiran, and N. M. Pindoriya, "Day-ahead Energy Management of a Microgrid with Battery Energy Storage Integration," *International Conference on Innovative Smart Grid Technologies, ISGT Asia 2018*, pp. 822–827, 2018.

[13] S. Van Der Kooij, P. Kempker, H. Van Den Berg, and S. Bhulai, "Optimal battery charging in smart grids with price forecasts," *2017 IEEE Power and Energy Society Innovative Smart Grid Technologies Conference, ISGT 2017*, 2017.

[14] Z. Wang and S. Wang, "Grid power peak shaving and valley filling using vehicle-to-grid systems," *IEEE Transactions on Power Delivery*, vol. 28, no. 3, pp. 1822–1829, 2013.

[15] S. Rahman and O. Hazim, "A generalized knowledge-based short-term load-forecasting technique," *IEEE Transactions on Power Systems*, vol. 8, no. 2, pp. 508–514, May 1993.

[16] Ying-Ying Cheng, P. P. Chan, and Zhi-Wei Qiu, "Random forest based ensemble system for short term load forecasting," in *2012 International Conference on Machine Learning and Cybernetics*. IEEE, Jul 2012, pp. 52–56.

[17] D. L. Marino, K. Amarasinghe, and M. Manic, "Building energy load forecasting using Deep Neural Networks," in *IECON 2016 - 42nd Annual Conference of the IEEE Industrial Electronics Society*. IEEE, Oct 2016, pp. 7046–7051.

[18] W. Kong *et al.*, "Short-Term Residential Load Forecasting Based on LSTM Recurrent Neural Network," *IEEE Transactions on Smart Grid*, vol. 10, no. 1, pp. 841–851, Jan 2019.

[19] A. Lahouar and J. Ben Hadj Slama, "Day-ahead load forecast using random forest and expert input selection," *Energy Conversion and Management*, 2015.

[20] S. Hochreiter and J. Schmidhuber, "Long Short-Term Memory," *Neural Computation*, vol. 9, no. 8, pp. 1735–1780, Nov 1997.

[21] A. Wallis, A. Haber, and S. Hauke, "Human knowledge and artificial intelligence in a forecast model for regenerative energies," in *16. Symposium Energieinnovation*, Graz, 2020.

[22] M. Kain, A. Haber, and M. Gaderer, "Security aspects of battery storages in parallel low-voltage grids," in *15. Symposium Energieinnovation*, 2018.

[23] O. Greil, J. Kainz, M. Kain, and A. Haber, "Optimizationmodel for charge and discharge strategies of PV-Storage-Systems," 2019, pp. 1–8.

Improving Energy Efficiency of Scientific Data Compression with Decision Trees

Michael Kuhn, Julius Plehn and Yevhen Alforov
University of Hamburg
Hamburg, Germany
Email: michael.kuhn@ovgu.de,juplehn@me.com,alforov@gmx.de

Thomas Ludwig
German Climate Computing Center
Hamburg, Germany
Email: ludwig@dkrz.de

***Abstract*—Scientific applications, simulations and large-scale experiments generate an ever increasing deluge of data. Due to the storage hardware not being able to keep pace with the amount of computational power, data reduction techniques have to be employed. Care has to be taken such that data reduction does not impact energy efficiency as it is an important cost factor for supercomputer systems and infrastructures. Data reduction techniques are highly data-specific and, therefore, unsuitable or inappropriate compression strategies can utilize more resources and energy than necessary. To that end, we propose a novel methodology for achieving on-the-fly intelligent decision making for energy-efficient data compression using machine learning. We have integrated a decision component into the Scientific Compression Library (SCIL) and show that, with appropriate training, our approach allows SCIL to select the most effective compression algorithms for a given data set without users having to provide additional information. This enables achieving compression ratios on par with manually selecting the optimal compression algorithm.**

Keywords–Data Compression; Energy Efficiency; Decision Tree.

I. Introduction

Even though the rate of improvement for processors has slowed down, the gap between the computational power of processors and other hardware components, such as main memory and storage, is still widening [1]. To keep pace, additional investments are necessary for storage hardware. This also results in additional costs for energy because new hardware used in computing systems requires additional power. Therefore, especially in data-intensive fields, costs for storage and energy are increasing. For instance, for each PetaByte (PB) of disk-based storage space, the German Climate Computing Center (Deutsches Klimarechenzentrum, DKRZ) has to pay investment costs of roughly 100,000 € and annual electricity costs of 3,680 €. For its 54 PiB storage system, this amounts to almost 200,000 € per year for electricity alone (one PB of storage needs 3 kW of power and 1 kWh of energy costs 0.14 €).

Data reduction techniques, such as compression, transformations and deduplication are straight-forward solutions to minimize the energy consumption of storage systems by reducing the amount of storage hardware required to store the same amount of data. However, data reduction itself can consume significant amounts of energy, potentially negating its beneficial effects on energy efficiency. While the energy efficiency of supercomputers should be increased, the impact on runtime performance should be minimal. A number of approaches and mechanisms to reduce energy consumption in supercomputers have been suggested at the different levels of computing systems. However, the impact of data reduction on High-Performance Computing (HPC) systems' energy efficiency remains largely unexplored, even though more and more HPC applications produce enormous volumes of data and data reduction techniques are increasingly adopted.

Developers of scientific software have a great interest in data reduction. However, to make best use of it, the used methods and algorithms have to be appropriate for their data sets and must be tuned to achieve optimal results. Additionally, decreasing runtime performance should be avoided both for performance reasons and for its impact on energy consumption. For these users, choosing a suitable compression algorithm is a technical decision that is difficult to make since the choice depends on the data set in question as well as the software and hardware environments. Data reduction schemes and options are highly data-specific and, therefore, our ultimate goal is to automatize the decision making process on behalf of the users. Poor manual choices can lead to low compression ratios, decreased performance and increases in energy consumption. In this paper, we are focusing on scientific applications in the context of HPC, where defining data reduction strategies with high performance and energy efficiency suitable for the generated deluge of scientific data is a challenging task.

Based on the methodology first introduced in [2], we define and extend mechanisms for intelligently selecting algorithms from a variety of state of the art reduction techniques with an emphasis on their energy consumption. In addition to employing machine learning to pick the most suitable compression strategy for a data set, users are able to specify additional criteria.

The remaining of the paper is structured as follows: In Section II, we give an overview of a common HPC I/O stack as it is used in earth system science. We introduce our framework for scientific data compression through high-level I/O interfaces in Section III. In Section IV, we present data collected for several compressors used to train our decision component. A detailed evaluation is performed in Section V using a real-world ecosystem simulation. After a review of related work in Section VI, we conclude in Section VII with a summary of our findings and describe our future work in Section VIII.

II. HPC I/O and Data Reduction

Applications typically use high-level I/O libraries to access data, which in turn uses I/O middleware to communicate with an underlying file system (see Figure 1). Two popular and common high-level I/O interfaces in the scientific community to access data in both serial and parallel manner are the *Hierarchical Data Format (HDF5)* and the *Network Common Data Form (NetCDF)*. They allow HPC applications written in various programming languages to manipulate and store data in a self-describing and portable way by using multidimensional arrays. Self-describing data formats contain a description of the file layout and are thus easy to share and distribute.

While NetCDF provides a convenient programming interface, the actual data format and I/O routines are implemented as part of HDF5. HDF5, in turn, uses the I/O implementation of the Message Passing Interface (MPI). MPI employs the I/O operations of the underlying parallel file system using optimized backends for a wide range of different file systems. In the end, I/O operations are posted by the file system to the underlying I/O driver. If the application performs data writing, it uses the high-level I/O library, and the data is going through the stack down until it is placed in the driver layer. A data read works in the opposite direction.

There are two main levels of the data path where data reduction mechanisms can be deployed: They are *system (low)* and *application (high)* levels, which each have different benefits and drawbacks. Data reduction usage on higher levels of HPC I/O stack is typically advantageous. Unlike low layers, it is possible to access and exploit additional meta information provided by the high-level I/O libraries. Different HPC applications for, e.g., climate change and weather forecasting, are using a common I/O stack, making it easier to employ application-level data reduction for them. Thus, usage of data reduction at the application level can be fine-tuned by taking application requirements and metadata into account. Among others, techniques such as deduplication, compression and transforms can be used on the application level. Moreover, as long as these techniques perform lossless data reduction, they can be deployed in a way that is transparent for users.

III. Energy-Efficient Data Compression

Based on these observations, we have extended the Scientific Compression Library (SCIL) to support energy-efficient data reduction by using machine learning approaches. SCIL already provides a rich set of features. In general, SCIL is a meta-compressor that aims to exploit knowledge on the application level [3]. The library should ultimately pick a suitable chain of algorithms satisfying the user's requirements. This is currently done based on the capabilities of the algorithms but has been extended by a decision component that can use different criteria, such as an energy-aware selection of the algorithms.

The overall architecture of using SCIL in scientific applications is depicted in Figure 1. Our main goal is providing the most appropriate data reduction strategy for a given scientific data set on the basis of semantical information and performance of algorithms. SCIL currently supports a wide range of lossless and lossy compression algorithms. Any application using the HDF5 data model can use SCIL via its HDF5 filter.

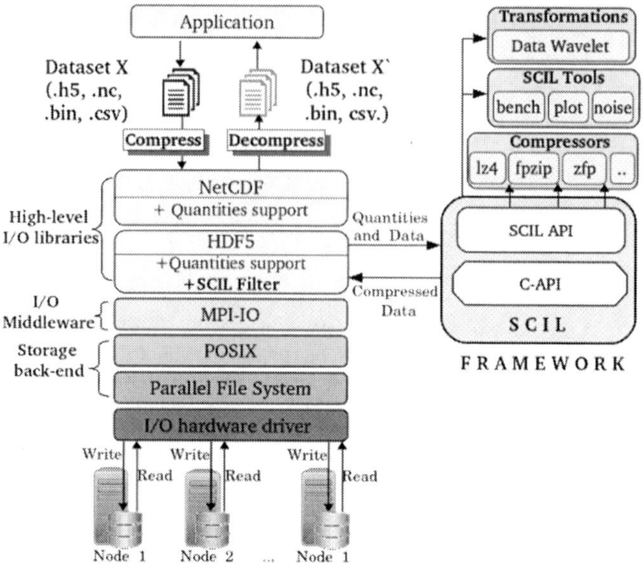

Figure 1. General architecture of Scientific Compression Library (SCIL) when integrated into the I/O workflow of scientific applications

A. Decision Component

Instead of relying solely on user-provided hints, we have extended SCIL with a decision component that takes into account information about the data's structure to provide improved data reduction capabilities. The decision component uses machine learning techniques to infer which compression algorithms and settings are best suited for a given data set. To provide enough information for the decision component to use, a separate training step is necessary. Currently, training is done separately from application runs by post-processing existing output data. The output data set is split up into its individual variables, which are then analyzed. For each variable, information about achievable compression ratios, processor utilization, energy consumption, etc. is collected. This is done for a wide range of data sets to provide the decision component with a sufficiently large pool of training data. The exact setup will be explained in more detail in Section V. Based on these measurements, it is possible to statically compute the appropriate compression algorithm for any given HDF5 variable. This information is then used to train the decision component, which currently makes use of decision trees but is planned to be extended with other techniques. In order to gain insights into the accuracy of those decision trees and to prevent overfitting, the data is split into a train and a test set. Those pre-processing steps are necessary in order to gather data that adds valuable information to the learning phase of the decision tree classifier.

The decision trees are created using the `DecisionTreeClassifier` component provided by scikit-learn, which produces a tree representation that is then parsed into a file that is usable in SCIL. Every time data is passed to SCIL, the decision component infers which compression algorithm and settings should be used before invoking its integrated compressors. The decision component's behavior can be separated into two distinct modes, as described below.

If the data's structure is known because it was part of the training set, the decision component can select the optimal compression algorithm and settings. This is currently done by comparing the data set's name but can be extended easily if necessary. This mode of operation is mainly useful for production runs of known applications. For instance, the decision component can be trained for a specific application and will be able to choose the best compressor for each subsequent run without the developers having to modify the application.

If the data's structure is unknown, the decision component will use machine learning techniques (currently, decision trees) to infer which compression algorithm and settings are best suited for the data set in question. The decision component will make use of information about the storage size, the number of elements, the data's dimensions, the data's type (float, double, etc.) and other factors. Moreover, it is possible to influence the decision based on whether energy efficiency, compression ratio or performance should be prioritized.

IV. Training of the Decision Component

Our current approach requires collecting several metrics, such as compression ratio, processor utilization and energy consumption for each supported compressor. Since the data's structure can heavily influence a compressor's behavior, these metrics are collected per data set. This collected information can then be employed in selection of power-aware data reduction techniques for a given data set. In the next paragraphs, we will gather the distinct performance and energy consumption characteristics of a wide range of compression algorithms through the use of HDF5 filters. Algorithms like LZ4 [4] and Zstandard [5] are fast and provide high throughputs. However, their compression ratios can be lower compared to slower algorithms that consume more energy. To exclude the influence of SCIL and the computation present in the actual applications, the training step uses output data generated by several real-world applications and experiments. The output data is split up into individual data sets and compressed using HDF5's `h5repack` utility. Overall, the evaluation has been conducted by repacking the data 10 times to obtain averaged metrics.

Hardware and Software Environment: In order to investigate the performance of data compression at the application level, we used a cluster, which operates with the parallel distributed file system Lustre. Since performance is not a priority, only a single node outfitted with two 2.80 GHz quad-core Intel Xeon X5560 processors and 12 GB of RAM was used to collect the required metrics. Due to this, the maximum throughput was limited to roughly 110 MB/s. To capture each compressor's energy consumption, the ArduPower wattmeter was used [6]. It is designed to simultaneously measure the power consumption of different components (e.g., motherboard, CPU, GPU and disks) inside computing systems even at very large scale. ArduPower provides 16 channels to monitor the power consumption with a variable sampling rate of 480–5,880 Hz.

Metrics: The main metrics in which we were interested are the Compression Ratio (CR) to quantify the data reduction, runtime of each algorithm to see how slow or fast is it, and consumed energy. We define compression ratio as $CR = \frac{\text{original size}}{\text{compressed size}}$.

Data Sets: For the evaluation of data reduction techniques, we used three data sets from different scientific domains:

1) ECOHAM: 17 GB data set produced by the 3-dimensional ecosystem model for the North Sea ECOHAM [7] (from climate science)
2) PETRA III: 14 GB data set of tomography experiments from PETRA III's P06 beamline (PCO 4000 detector) [8] (from photon science)
3) ECHAM: 4 GB data set produced by the atmospheric model ECHAM [9] (from climate science)

Evaluated Techniques: To perform the reduction of data sets, different HDF5 compression filters have been leveraged. As part of our experimental evaluation, we have compared the following algorithms commonly used in HPC. While others, such as DEFLATE, LZMA or xz also offer high compression ratios, their performance is typically not sufficient for HPC workloads.

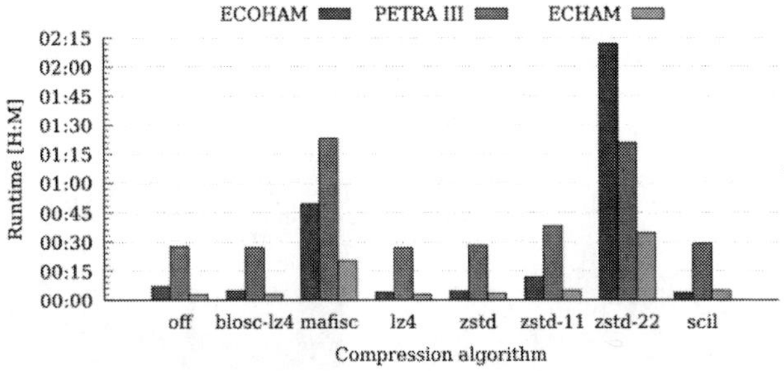

Figure 2. Runtime of evaluated compressors

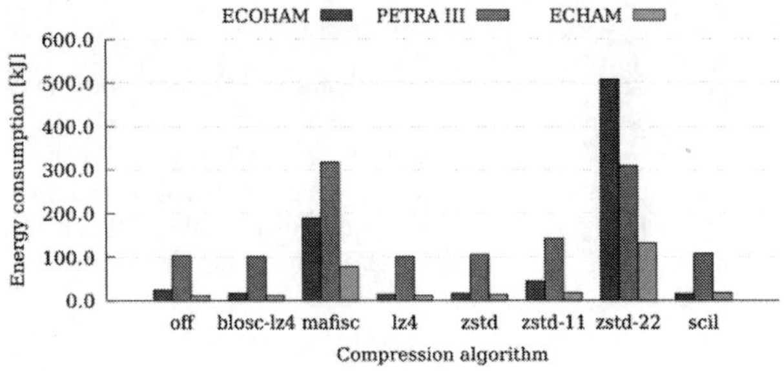

Figure 3. Average energy consumption depending on the HDF5 filter used for data compression

Note that the following results only include compression as decompression is usually much faster and thus negligible.

off: No filtering is applied. This represents the baseline. *blosc*: The Blosc meta-compressor using the LZ4 compressor. Additionally, Blosc's shuffle pre-conditioner was used. *mafisc*: The MAFISC compression algorithm that uses several pre-conditioners and the LZMA compressor. *lz4*: The LZ4 compression algorithm using its default acceleration factor. *zstd*: The Zstandard compression algorithm using its default aggression parameter. The *zstd-11* and *zstd-22* variants represent Zstandard with aggression parameters of 11 and 22, respectively. *scil*: SCIL's LZ4 compressor with some pre-conditioners.

Figure 2 shows that the runtimes vary wildly depending on the internal structure of the datasets. The consumed energy and obtained compression ratios for the ECOHAM, PETRA III and ECHAM data sets are plotted in Figures 3 and 4, respectively. It can be seen that different configurations achieve comparable

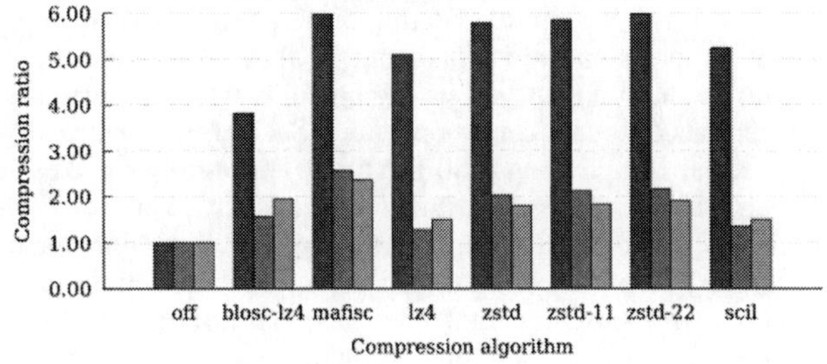

Figure 4. Average compression ratios depending on the HDF5 filter used for data compression

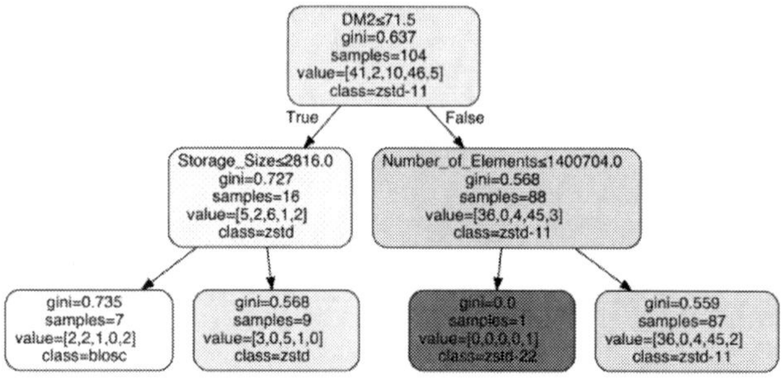

Figure 5. Decision component trained with ECOHAM data, optimized for maximal compression ratio per time

compression ratios with significantly different runtimes and energy consumptions (for instance, compare mafisc and zstd for the ECOHAM dataset). It is, therefore, necessary to select the compression algorithm intelligently to avoid wasting performance and energy. For a detailed analysis of the results for the ECOHAM and PETRA III datasets, please refer to [2].

V. Evaluation

Compressors behave very differently depending on the data structure. Based on the results and data obtained in the previous section, we have trained the decision component using two different sets of training data and will run ECOHAM using our SCIL HDF5 plugin to evaluate our approach. All relevant code and data for this paper are available at [10].

ECOHAM data used as training data: This configuration represents the case that the application in question is known and has been run before on the system. The decision component has knowledge about this particular application's data and can take informed decisions regarding data reduction. However, only a random subset consisting of 75 % of ECOHAM's output data is used for training to make sure there is a degree of uncertainty left. This uncertainty could correspond to updated versions of the application or slightly changed output structure.

ECHAM data used as training data: This configuration represents the case that a new application is run on a system and the decision component has to use information gathered from other applications to try to compress the application's data as best as possible. In this case, the decision component will try to map decisions that make sense for other data sets to the current application's data.

Moreover, we will look at two different optimization targets for the decision component, which correspond to different use cases: First, we optimized for minimal energy consumption per compression ratio. This strategy allows shrinking the data with the least amount of energy possible, which is typically of importance to data center operators. Second, we optimized for maximal compression ratio per time. This strategy makes sure that performance is not degraded excessively, which is usually one of the main concerns of data center users. For instance, Figure 5 shows a configuration using ECOHAM training data with a decision tree that has been optimized for maximal compression ratio per time. As can be seen, a multitude of metrics are taken into account, including the array dimensions, the data set's storage size, as well as the number of elements. In addition to these metrics, the size of each dimension and

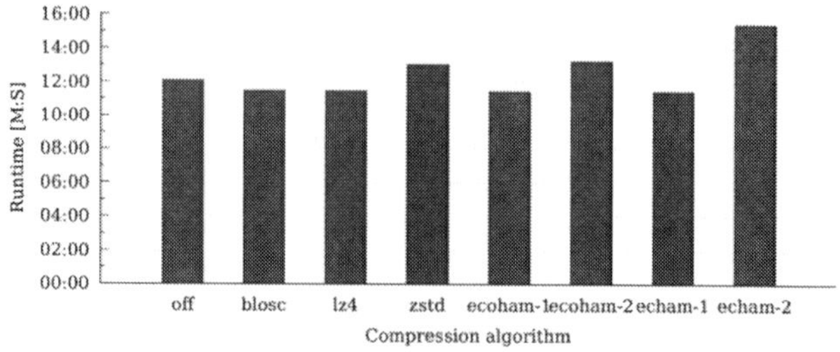

Figure 6. Runtime of evaluated compressors

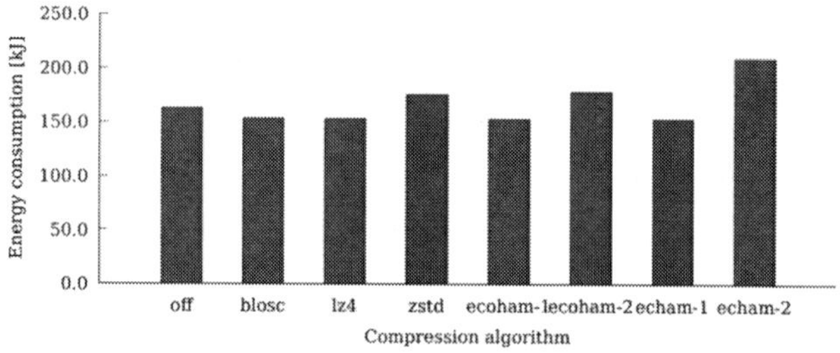

Figure 7. Energy consumption of evaluated compressors

information about data types can be used. In this specific case, the decision tree makes sure that a mix of the Blosc and Zstandard compressors are used. Moreover, Zstandard's compression level is adapted for maximum effect.

To compare our approach to static approaches, we have chosen to run ECOHAM with the most important compressors in static mode (that is, all data is compressed with the selected algorithm) and four of our decision trees. It is important to note that, although only Blosc, LZ4 and Zstandard are used as static approaches, the decision trees have access to all of SCIL's compressors, which also include different aggression parameters for Zstandard. Figures 6, 7 and 8 show the runtime, energy consumption and compression ratio of ECOHAM when run with the following configurations.

off: No HDF5 filter is used and, thus, no compression is performed. *blosc*: SCIL's HDF5 filter is configured to compress ECOHAM's data using Blosc. *lz4*: SCIL's HDF5 filter is configured to compress ECOHAM's data using LZ4. *zstd*: SCIL's HDF5 filter is configured to compress ECOHAM's data using Zstandard. *ecoham-1*: The decision component has been trained with ECOHAM data and is optimizing for minimal energy consumption per compression ratio. *ecoham-2*: The decision component has been trained with ECOHAM data and is optimizing for maximal compression ratio per time. *echam-1*: The decision component has been trained with ECHAM data and is optimizing for minimal energy consumption per compression ratio. *echam-2*: The decision component has been trained with ECHAM data and is optimizing for maximal compression ratio per time.

As can be seen in Figures 6 and 7, fast compression algorithms, such as LZ4 reduce runtime and energy consumption by causing less data being written to the file system and thus slightly speeding up the whole application. The same is true for the other light-weight algorithm Blosc and SCIL's default configuration. However, slower algorithms, such as Zstandard

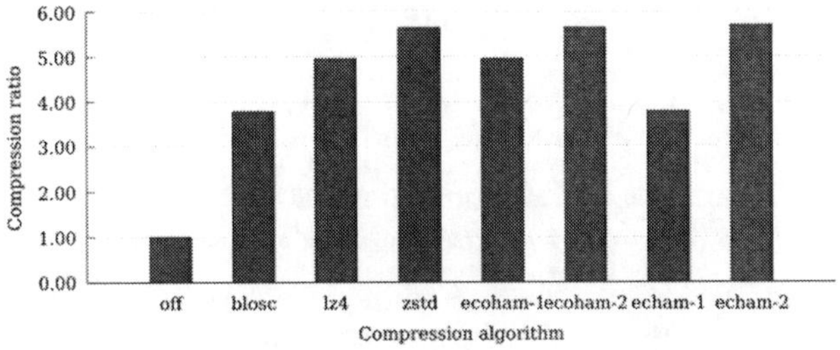

Figure 8. Compression ratio of evaluated compressors

have the opposite effect and cause increases in runtime and energy consumption. Most importantly, the decision component correctly uses energy-efficient algorithms when optimizing for minimal energy consumption per compression ratio, which can be seen in the *ecoham-1* and *echam-1* configurations. In the *ecoham-1* case, ECOHAM's data structure is known and this information can be used to effectively reduce the amount of data. In the *echam-1* case, however, the decision component only has knowledge about ECHAM's data structures but is still able to choose appropriate compressors. The *ecoham-2* and *echam-2* configurations both increase the runtime and energy consumption. This is expected since, in contrast to the previous cases, the optimization for maximal compression ratio per time puts more emphasis on data reduction instead of low energy consumption. Since the data structure is not known in the *echam-2* case, the decision component's choices are not as effective as in the *ecoham-1* case and increase runtime and energy consumption to a higher degree. Overall, the results show that the decision component can be used both for known as well as unknown applications. However, if the application's data structure are known, better decisions can be made.

This can also be seen in Figure 8, which illustrates the compression ratios achieved by all configurations. As expected, Blosc's compression ratio is the lowest (3.79), followed by LZ4 (4.95) and Zstandard (5.64). For *ecoham-1* and *ecoham-2*, the decision component has knowledge about ECOHAM's data structures and chooses the optimal algorithms for both optimization targets. When optimizing for low energy consumption (*ecoham-1*), the decision component favors LZ4 and achieves a compression ratio of 4.95. When a higher compression ratio is preferred (*ecoham-2*), Zstandard is chosen most of the time, leading to a compression ratio of 5.64. When the decision component has been training with ECHAM's data, the results are different: For *echam-1*, the decision component chooses Blosc most of the time, which achieves the goal of reaching a low energy consumption but boasts a lower compression ratio of 3.79. For *echam-2*, however, the decision component does end up using significantly more energy than *ecoham-2* but also provides a higher compression ratio of 5.69. This is due to the fact that the decision component also chooses Zstandard with higher aggression parameters.

In order to put the possible energy savings of this approach into perspective, the initially required energy consumption of the training has to be compared. For every variable existing in the data set, all available compression algorithms and selected compression levels are applied. In case of ECOHAM, 1,953 tests have been performed, resulting in 10,633 kJ being consumed. For ECHAM, 833 tests were necessary with a total energy consumption of 1,169 kJ. A comparison of the energy consumptions from Figure 7 shows that our approach can save roughly 10 kJ per run in contrast to using no compression. Even though these savings appear small, the presented runs were relatively short with 11–12 minutes. Production runs of these applications typically take several days or weeks and are repeated many times for comparison purposes. Therefore, significant cost savings are possible even though initial training costs appear high. Moreover, training costs can be decreased by eliminating algorithms and settings that prove to be inefficient.

VI. Related Work

The results by Welton et al. [11] show that the achievable throughput is highly dependent on the chosen algorithm and data properties because slow algorithms or incompressible data can decrease throughput significantly. One way to compensate for this drawback is to implement these algorithms in hardware. Abdelfattah et al. [12] have implemented gzip on Field-Programmable Gate Arrays (FPGAs) using OpenCL, which offers higher throughput and a better performance-per-watt ratio. Intel's QuickAssist technology can also be used to lower the total cost of ownership by executing popular encryption and compression algorithms in hardware, as shown by Hu et al. [13]. However, all of these approaches still require developers to manually select a compression algorithm and settings. Inappropriate choices can lead to suboptimal performance and compression ratios. It is, therefore, important to foresee which reduction method will produce the best results. For example, Chen et al. [14] present a decision algorithm for MapReduce users to decide whether to use compression or not. They reported that compression provides up to 60 % energy savings for some jobs.

Machine learning techniques (especially Deep Learning) are being increasingly used to compress images and videos, as shown by Liu et al. [15]. They show that new approaches based on deep networks can produce comparable results to traditional coding approaches. Neural networks have also been used by Park et al. [16] to compress data gathered by Internet of Things devices in a lossy fashion. Rippel et al. [17] created a machine-learning-based approach for lossy image compression that outperforms traditional approaches, such as JPEG, JPEG 2000 and even WebP. A similar approach has been taken by Toderici et al. [18], who have created a compressor based on Recurrent Neural Network (RNNs) that also outperforms the traditional lossy JPEG compressor. The same is possible for lossless image compression, as shown by Mentzer et al. [19]: The proposed image compression system L3C outperforms PNG, WebP and JPEG2000. Machine learning can also be used for indirect space savings: Kraska et al. [20] have used machine learning to replace traditional data structures for b-trees, hash maps and bloom filters, which allowed reducing the amount of data needed by these data structures while simultaneously delivering competitive performance. However, in these cases, the actual compression is replaced using machine learning. This has the downside of not being able to perform lossless compression, which our approach can achieve.

VII. Conclusion

In this work, we have analyzed whether it is possible to automatically and intelligently pick compression algorithms for a given data set by making use of machine learning techniques. Our results show that the amount of data which can be saved after using reduction techniques like compression heavily depends on the structure of data. Preconditioners, algorithms and settings might work well for one data set, but they might increase energy consumption for others. The preliminary results obtained during the training step have been taken into account when designing and implementing the decision unit for intelligent algorithms selection in SCIL. We have used fine-grained per-variable analyses to identify the optimal compression strategies for three different data sets and used this data to train the decision component for our real-world evaluation. We could demonstrate that the decision component is able to choose appropriate compressors for both known and unknown applications, which can be tuned for energy efficiency or compression ratio. Without providing additional information, the decision component was able to achieve satisfactory compression ratios without increases in energy consumption. Moreover, by changing the optimization strategy of the decision trees to allow slight increases in energy consumption, we could significantly boost compression ratios.

VIII. Future Work

Training currently has to be performed in a separate step. In the future, we envision training data collection to be more tightly integrated with production runs of applications. For instance, a specialized training mode of SCIL's HDF5 filter could be used to capture and analyze applications' output data during regular runs. In order not to influence application performance negatively, selected data samples could be sent to a training service that then takes care of analyzing it in more detail using a wide range of compression algorithms. For instance, a preloadable library, which intercepts calls to HDF5 and integrates filters into applications could offer this functionality in a transparent way. This would also allow us to extend our analysis to more compression algorithms and datasets. Since training is a manual process at the moment, we have focused on algorithms commonly used in HPC for now.

Additionally, the current interface used by HDF5 filters is too limiting to fully exploit all possibilities offered by our decision component. For instance, since HDF5 filters operate on opaque buffers, it is not easily possible to access single data points. However, this could be used gather further information about data variance, such as maximum and minimum values, which could be used to further tune compressor behavior. We will also experiment with chains of compressors. Applying multiple compressors in the correct order can lead to additional space savings. However, figuring out the order and suitable compressors is a combinatorial problem that is not easy to solve manually. Therefore, we want to extend the decision component to take this fact into account and predict chains of compressors instead of singles ones.

Acknowledgment

This work was supported in part by the BigStorage project, funded by the European Union under the Marie Skłodowska-Curie Actions (H2020-MSCA-ITN-2014-642963). We would also like to thank Anastasiia Novikova for her support with using the SCIL library and André Rothkirch from DESY for providing us with access to parts of their data.

References

[1] R. Chen, Z. Shao, and T. Li, "Bridging the I/O performance gap for big data workloads: A new NVDIMM-based approach," in 49th Annual IEEE/ACM International Symposium on Microarchitecture, MICRO 2016, Taipei, Taiwan, October 15-19, 2016, pp. 9:1–9:12. [Online]. Available: https://doi.org/10.1109/MICRO.2016.7783712

[2] Y. Alforov, T. Ludwig, A. Novikova, M. Kuhn, and J. M. Kunkel, "Towards Green Scientific Data Compression Through High-Level I/O Interfaces," in 30th International Symposium on Computer Architecture and High Performance Computing, SBAC-PAD 2018, Lyon, France, September 24-27, 2018, pp. 209–216. [Online]. Available: https://doi.org/10.1109/CAHPC.2018.8645921

[3] J. M. Kunkel, A. Novikova, E. Betke, and A. Schaare, "Toward Decoupling the Selection of Compression Algorithms from Quality Constraints," in High Performance Computing - ISC High Performance 2017 International Workshops, DRBSD, ExaComm, HCPM, HPC-IODC, IWOPH, IXPUG, P^3MA, VHPC, Visualization at Scale, WOPSSS, Frankfurt, Germany, June 18-22, 2017, Revised Selected Papers, pp. 3–14. [Online]. Available: https://doi.org/10.1007/978-3-319-67630-2_1

[4] Y. Collet, "LZ4," https://lz4.github.io/lz4/, 2020, retrieved: September, 2020.

[5] Facebook, "Zstandard," https://facebook.github.io/zstd/, 2020, retrieved: September, 2020.

[6] M. F. Dolz, M. R. Heidari, M. Kuhn, T. Ludwig, and G. Fabregat, "ArduPower: A low-cost wattmeter to improve energy efficiency of HPC applications," in Sixth International Green and Sustainable Computing Conference, IGSC 2015, Las Vegas, NV, USA, December 14-16, 2015, pp. 1–8. [Online]. Available: https://doi.org/10.1109/IGCC.2015.7393692

[7] F. Große et al., "Looking beyond stratification: a model-based analysis of the biological drivers of oxygen depletion in the North Sea," Biogeosciences Discussions, 2015, pp. 2511–2535, retrieved: April, 2020. [Online]. Available: http://www.biogeosciences-discuss.net/12/12543/2015/bgd-12-12543-2015.pdf

[8] DESY, "PETRA III," http://petra3.desy.de/index_eng.html, 2015, retrieved: April, 2020.

[9] E. Roeckner et al., "The atmospheric general circulation model ECHAM 5," Max Planck Institute for Meteorology, 2003.

[10] M. Kuhn, J. Plehn, and Y. Alforov, "Supplementary Material for Improving Energy Efficiency of Scientific Data Compression with Decision Trees," https://github.com/wr-hamburg/energy2020-compression, 2020, retrieved: September, 2020.

[11] B. Welton et al., "Improving I/O Forwarding Throughput with Data Compression," in 2011 IEEE International Conference on Cluster Computing (CLUSTER), Austin, TX, USA, September 26-30, 2011, pp. 438–445. [Online]. Available: https://doi.org/10.1109/CLUSTER.2011.80

[12] M. S. Abdelfattah, A. Hagiescu, and D. Singh, "Gzip on a chip: high performance lossless data compression on FPGAs using OpenCL," in Proceedings of the International Workshop on OpenCL, IWOCL 2013 & 2014, May 13-14, 2013, Georgia Tech, Atlanta, GA, USA / Bristol, UK, May 12-13, 2014, pp. 4:1–4:9. [Online]. Available: https://doi.org/10.1145/2664666.2664670

[13] X. Hu et al., "QTLS: high-performance TLS asynchronous offload framework with Intel QuickAssist technology," in Proceedings of the 24th ACM SIGPLAN Symposium on Principles and Practice of Parallel Programming, PPoPP 2019, Washington, DC, USA, February 16-20, 2019, pp. 158–172. [Online]. Available: https://doi.org/10.1145/3293883.3295705

[14] Y. Chen, A. Ganapathi, and R. H. Katz, "To compress or not to compress – compute vs. IO tradeoffs for MapReduce energy efficiency," in Proceedings of the 1st ACM SIGCOMM Workshop on Green Networking 2010, New Delhi, India, August 30, 2010, pp. 23–28. [Online]. Available: https://doi.org/10.1145/1851290.1851296

[15] D. Liu, Y. Li, J. Lin, H. Li, and F. Wu, "Deep Learning-Based Video Coding: A Review and A Case Study," CoRR, vol. abs/1904.12462, 2019. [Online]. Available: http://arxiv.org/abs/1904.12462

[16] J. Park, H. Park, and Y. Choi, "Data compression and prediction using machine learning for industrial IoT," in 2018 International Conference on Information Networking, ICOIN 2018, Chiang Mai, Thailand, January 10-12, 2018, pp. 818–820. [Online]. Available: https://doi.org/10.1109/ICOIN.2018.8343232

[17] O. Rippel and L. D. Bourdev, "Real-Time Adaptive Image Compression," in Proceedings of the 34th International Conference on Machine Learning, ICML 2017, Sydney, NSW, Australia, 6-11 August 2017, pp. 2922–2930, retrieved: April, 2020. [Online]. Available: http://proceedings.mlr.press/v70/rippel17a.html

[18] G. Toderici et al., "Full Resolution Image Compression with Recurrent Neural Networks," in 2017 IEEE Conference on Computer Vision and Pattern Recognition, CVPR 2017, Honolulu, HI, USA, July 21-26, 2017, pp. 5435–5443. [Online]. Available: https://doi.org/10.1109/CVPR.2017.577

[19] F. Mentzer, E. Agustsson, M. Tschannen, R. Timofte, and L. V. Gool, "Practical Full Resolution Learned Lossless Image Compression," CoRR, vol. abs/1811.12817, 2018, retrieved: April, 2020. [Online]. Available: http://arxiv.org/abs/1811.12817

[20] T. Kraska, A. Beutel, E. H. Chi, J. Dean, and N. Polyzotis, "The Case for Learned Index Structures," in Proceedings of the 2018 International Conference on Management of Data, SIGMOD Conference 2018, Houston, TX, USA, June 10-15, 2018, pp. 489–504. [Online]. Available: https://doi.org/10.1145/3183713.3196909

ArduPower v2: Open and Modular Power Measurement for HPC Components

Daniel Bremer, Michael Kuhn and Mohammad Reza Heidari
Universität Hamburg
Hamburg, Germany
Email: {daniel.bremer, michael.kuhn, heidari}@informatik.uni-hamburg.de

Abstract—**Accurate power measurement is a prerequisite to energy optimization in High Performance Computing (HPC) systems. Fine temporal and spatial power profiling is also required to capture fast power fluctuations that can point to inefficiencies in applications, libraries or the operating system. Moreover, flexible deployment of powermeters is an important factor to apply power measurement in computing systems without breaking into the system circuitry or compromising the system warranty. However, it is usually a challenge to find a comprehensive solution in the market that meets all the requirements to study energy efficiency in HPC systems. In response to the various research needs, ArduPower aims at a modular internal wattmeter platform. It enables flexible deployment in computing systems, renders accurate power monitoring with a high sampling rate at component level and offers an on-the-fly power consumption analysis for given components. Its open design and low price lend it to an affordable option for fast deployment and further customization for specific needs. Our evaluations show that ArduPower is capable of delivering competitive results with commercial solutions which allows us to identify application phases within popular benchmark programs. This information can be exploited to tune for higher computational performance or energy efficiency in computing systems.**

Keywords–Internal Wattmeter; Power Consumption; Energy Efficiency; ArduPower; Modular Architecture.

I. Introduction

In many fields in research and engineering, High-Performance Computing (HPC) is required to process data or simulate processes or structures. While computation clusters were rated and evaluated by their performance and Floating Point Operations per Second (FLOPS) for a long time, today's evaluation processes also consider operating costs. Besides maintenance, a huge factor for these is energy consumption. This shift is also shown with the introduction of the Green500 list in 2009 which, opposing to the TOP500 list, ranks systems using their energy efficiency by considering watts per GFLOPS. While the TOP500 is dominated by Graphics Processing Unit (GPU) powered systems, the Green500 list places an ARM-based CPU in the top position, which only reaches place 159 in the TOP500 [1]. The ArduPower v2 platform, introduced in this paper, aims to enable cheap, plug and play power measurement for many systems, allowing for optimizations in energy efficiency in regards to hardware selection, as well as programming techniques.

Measurement of power consumption is done with wattmeters. The formula for electric power P is given by $P = U * I$, where U is voltage and I is the current. Normally, the voltage on a power delivery line is known relatively well and should be in a specified range, e.g., the ATX standard specifies $\pm 5\,\%$ for positive lines. Acquiring a value can be easily done by either consulting the standard's documentation or by probing with a voltmeter, which is connected in parallel, or an Analog-Digital-Converter.

The current, on the other hand, fluctuates with the current workload of the consumer circuit and often is an unknown value. To measure its value, most often Hall-effect sensors are used to determine the strength of the magnetic field induced by a current passing a conductor, which directly correlates to the current value. Measuring current can be done with a clamp meter, which is placed around a singular conductor and captures the induced magnetic field strength, or by placing an ammeter in series with the consumer.

Introducing a wattmeter for energy analysis can be done externally or internally. External wattmeters, such as ZES Zimmer LMG450 or similar, monitor power consumption between the power outlet and a server's power supply. Readings represent the power draw of a system as a whole, but only allow an overall estimation of real load. Factors offsetting the values are the efficiency of the power supply or sporadic, non-application related workloads. A detailed analysis often is hard and power fluctuations of single components can not be identified.

Internal wattmeters aim to provide power draw readings on a component level, differentiating between energy used by processors, mainboard, hard disk drives, etc. A finer level of analysis is possible as different application behaviors like computation bound or memory bound show on their corresponding channels and in the specific energy profiles. Results also can be transformed for system-level analysis by summing up all channels. Combining both variants, external and internal, the efficiency of power supply units can be evaluated. With ArduPower v2 an internal wattmeter is introduced. Evolved from the ArduPower platform, it delivers more features while maintaining the already proven concept.

II. Uses of Power Consumption Analysis

Building a computation cluster is a complicated task requiring optimization of multiple features. A naive approach to this can be simply to optimize available computational resources, like FLOPS. This might work for building a capable general purpose cluster, but often systems face specific workloads repeatedly, which might be a reference for tuning. For instance, systems running ligand configuration simulations often benefit from high core counts, as the tasks are worked on in a batch-like manner, therefore parallel and independent simulations are done. Engineering tasks, however, are regularly calculations-heavy and benefit from faster floating point arithmetics, that are found on higher clock-frequency CPU models.

Besides clock speeds and thread counts, there are also architectural differences. Generally, CPU clusters are suited for every computational problem, but, in recent years, development of new programming techniques and hardware architectures pushed the usage of GPUs and accelerator cards, which provide higher FLOPS than regular CPUs. Also, ARM's technology emerged, providing low-powered CPUs with high energy efficiency. Besides the hardware cost for a new system, energy cost accumulates over the time of usage, reaching a significant portion of the initial costs. HPC systems have a typical lifetime of around 5–6 years and energy costs over this period of time can easily add another 20 % of total procurement costs. These costs include electricity for powering the machine as well as costs for cooling. Due to the high costs of supercomputers, reducing energy costs even by small amounts can result in huge cost savings. In periods of inactivity, servers are typically not shut down but left idling, during which they still consume significant amounts of energy. While external and non-critical devices like GPUs can be turned off completely, CPUs can typically only be put into power-saving modes.

Power consumption analysis allows evaluating components on an energy level. When comparing configurations, a standardized set of programs can be run to model general cluster usage and energy consumption can be tracked. If the results show the efficiency of GPUs over CPUs for the desired use case, such a more specialized cluster could be taken into consideration. If computation speed can be sacrificed for energy efficiency, ARM-based CPUs could offer an alternative over regular x86 CPUs. Evaluating feature changes can be done as well. While GPUs often perform faster floating point arithmetics than CPUs, modern chips provide vectorization with AVX512 and therefore can heavily boost performance, and in some cases even outperform GPUs [2].

In operation, power consumption monitoring can be used to allocate resources to users. Traditional job managers allocate available nodes by considering real-world time slots, CPU time or a type of credit. Instead of providing a 6-hour time slot on a machine, the job manager could implement a power limit that could be set over a certain time span. This would motivate users to use more energy-efficient code, to maximize runtimes of programs, and be mindful when planning jobs. Managers like SLURM [3] support power management natively, which can manage nodes power caps and allows for accounting of consumed energy.

III. RELATED WORK

Reliable power measurement is a prerequisite to energy efficiency as the key optimization goal in HPC systems. Despite considerable efforts, there is a wide spectrum of applications that yet demand more flexible power measurement in HPC systems. Power monitoring solutions are usually dictated by the applications' power profiles, which affect the required sampling rate, measurement accuracy, spatial granularity, scalability and instrumentation. More often than not, it is highly desirable to leverage existing flexibility that lends itself to a variety of applications and customizations [4].

External power measurement is a common practice to study power consumption at node level in computing systems. Power Supply Units (PSUs) and Power Distribution Units (PDUs) often provide the overall power consumption of a computing node via the Intelligent Platform Management Interface (IPMI) with a relatively low precision and sampling rate [5]. However, they fail to capture power traces that last only a few seconds [6]. Professional AC wattmeters, such as ZES Zimmer, offer accurate average power measurement externally at node level but their low sampling rate cannot reflect fast power fluctuations for detailed analysis [7]. The external power measurement approach usually suffers from a low spatial granularity that is an obstacle in the power analysis at component level.

Internal power measurement is, however, a way forward to achieve more flexible power monitoring as it usually offers higher spatial and temporal granularities. PowerInsight, designed by Sandia National Laboratories and Penguin Computing, is an internal wattmeter with 15 channels and provides instantaneous DC power measurement at a maximum total sampling rate of 1,000 Sa/s [8]. It offers a probe-based monitoring connected to a BeagleBone for data collection and forwarding. Its design supports standard power plugs as well as PCI-e risers to monitor devices such as GPUs. PowerMon2 is also an internal wattmeter that provides instantaneous power measurement at a maximum sampling rate of 1,024 Sa/s and supports up to 8 DC channels [9]. PowerPack is another internal power monitoring system that supports dynamic power management configurations [10]. It provides component-based power monitoring through a dedicated data acquisition mechanism with support for multicore processors and power-performance analysis affected by the DVFS mechanism.

New HPC systems embed vendors' proprietary solutions to estimate energy consumption system-wide from node level to the component level. As a prominent example, the High Definition Energy Efficiency Monitoring (HDEEM) is a sophisticated approach, designed by Technische Universität Dresden and Bull S. A. S. [11]. Its goal is a fine-grained power measurement enabled by a sampling rate of 8,000 Sa/s over a fine spatial granularity, e.g., for per-CPU measurements. Cray XC30 system series also offers an integrated power measurement infrastructure at component level that includes blade and GPU measurements at 10 Sa/s [12]. IBM applies a similar integrated circuits approach to Power7 processors to monitor power consumption at a component level [13].

While the PowerInsight [8] approach is similar to ArduPower in regard to deployment, the feature sets differ vastly. Featuring a BeagleBone board, PowerInsight has potential to directly store captured data on a drive or stream it to another endpoint via Ethernet, but is limited in terms of connectivity, as the internal 7-channel ADC only is referenced to 1.8 V, making it crucial to implement a dedicated ADC for 5 V references sensors on the "Power Cape" board. ArduPower, on the other hand, is better geared towards customization, featuring an Arduino Mega. This provides 16 native 5 V referenced ADC channels and more connectivity via GPIO pins, allowing implementation of SPI or serial buses for further devices, not just for power monitoring, but also for arbitrary sensors like thermal probes. Providing a richer interface for sensors, the automatic configuration feature enables plug and play installation, leading to fast deployment.

IV. THE ARDUPOWER PLATFORM

The first implementation of ArduPower was presented by Dolz et al. and was based on a custom shield for the Arduino Mega platform, providing 16 inputs for power lines [14]. After a configuration process, the system could be connected to a power wiring harness inside a computation node, by cutting wires and rerouting them to the shield, and data was sent with an 8-bit serial interface to the monitoring node.

While the wattmeter itself worked flawlessly, the system required much manual configuration. Version 2 re-implements the idea of using Hall-effect sensors for current measurement and introduces new features, such as an automatic configuration of the device and an improved serial protocol. To enable fast setup and modularity, the platform was split into a collector unit, consisting of an Arduino Mega 2560 and a shield, which is used for internal line routing and automatic configuration circuitry. Moreover, the Hall-effect sensors were moved onto dedicated probe devices to modularize the approach.

Probes are designed to be female-male extension cables equipped with Hall-effect sensors. This allows integration into the existing wiring harness with no further manipulation. All installed probes then need to be connected with the collector unit by cable to be forwarded to the monitoring node.

A. *Hardware*

In the following, ArduPower v2's hardware design will be described in detail, which is illustrated in Figure 1.

Probes: Voltages on internal power lines are known to be either 3.3 V, 5 V or 12 V with an error of 5 %, as given by the ATX specification. Components with high power consumption like CPUs or GPUs normally use 12 V for power delivery, while Solid-State Drives (SSDs) and USB devices do not consume as much power and, therefore, mainly rely on lower voltages. The reason to use different voltage levels is to keep currents low. Both voltage and current are a linear factor in electric power, therefore, it is possible to transport 60 W of power with 5 V and 12 A or by using a voltage of 12 V and 5 A of current. The reasons to choose higher voltages are smaller wire diameters, reduced magnetic fields (induced by the Hall-effect) and safety.

As the voltages are known due to standardization, probes need to be put in series with the power supply and power consuming component. For simple connection, a probe is designed similar to an extension cable with a small circuit board interrupting the power carrying line. This circuit then contains an ACS723 Hall-effect sensor by Allegro and connections to provide power to the probe and return data to the collector unit. To allow for better accuracy, the bidirectional, 40 A-rated sensor for high power probes (e.g. PCI-e power to GPUs) and the bidirectional, 20 A variant for lower-powered devices were chosen. Connection pins on the probe are used for 5 V power in and GND for power delivery and for an analog signal from the ACS723 sensor which represents the current, a type pin to determine the used ACS723 variant and a pin carrying the line voltage to the ArduPower collector unit, that is used by the automatic configuration.

Shield: Arduino shields are circuit boards with a layout that can directly be stacked onto a device's pins and allow for hot-pluggable reconfiguration of functionalities. ArduPower's version 1 directly mounted the Hall-effect to this board, while version 2 moved the sensors to the probes and used the shield for internal wire routing. It also features an analog multiplexer chip, a voltage divider and connection pins to interface with the probes.

ArduPower v2 implements an automatic configuration that collects information on the connected probes to determine their type and the voltage of the measured line. As for this, an analog voltage must be digitalized and each voltage has to be sensed by the Arduino's Analog-Digital-Converter (ADC). But, as the

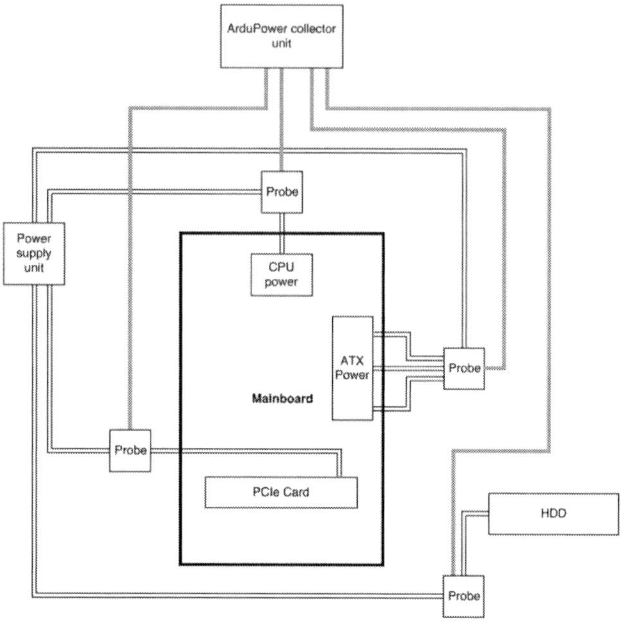

Figure 1. Implementing ArduPower v2 is done by placing probes between the power supply unit and each different consumer components, such as CPU, GPU or storage devices. Each probe then is connected to the ArduPower collector unit, which itself is connected to a monitoring node.

output signal from the probes also is an analog signal, this, too, has to be sent to the ADC, which maxes out the Arduino's capabilities, as the ADC only provides 16 inputs. An external multiplexer was built onto the shield to allow for input of 32 analog signals, enabling sensing all probed lines' voltages and the output values of the sensors.

Another crucial step identifying the line voltages on the probes was the introduction of a voltage divider. The Arduino Mega has a rated pin voltage of 5 V, while computer power lines carry up to 12 V and therefore would destroy the device. To circumvent this, a voltage divider was set up, remapping the 0 V to 12 V range to a 0 V to 5 V range, which can be read by the ADC.

Automatic Configuration: Previous versions of ArduPower required a configuration file for voltage information like probe voltage. Version 2 implements an automatic configuration feature. By forwarding the voltage of the probed line or another set voltage between 5 V and 12 V from the probe to the collector unit, the voltage can be automatically identified and used for power calculation. The user only must identify the measured line, e.g., whether a 12 V line is connected to an EPS connector, providing power to a CPU or a PCI device, like a GPU. This feature requires a combination of hardware and software. While a voltage divider and an additional analog multiplexer were introduced on the ArduPower shield, the control of the multiplexer is realized with software. Voltage measurement is handled by the Arduino's ADC with the `analogread` method, while voltage identification is done in software, as the voltage divider shifts the measured voltages into a 0 V to 5 V range.

Further optimization is done by only considering connected probes when performing a measurement step. Analog-digital-conversion is a relatively slow operation, therefore, only a rate of 9,600 Hz can be achieved, as each ADC cycle takes about 110 μs. With each start of a measurement, connected probes are stored and only the corresponding ports are then covered in the measurement cycle.

Deployment: Deploying the ArduPower wattmeter is a process of installing the probes between the components that should be power-traced and the PSU. For this, standard power headers such as ATX, PCI-e or EPS and a probe featuring the corresponding connector are required. Physically, no modifications have to be done other than adding the probes to the existing wire harness and routing wires from the probes to the collector unit and from this unit to an available USB port.

Server style mainboards are often fed with only a 12 V line providing most of the power, which then is transformed to the respective operating voltage of the component. Such boards are currently not supported due to the limitation of space in such chassis.
Larger scale deployments can be realized by installing probes into multiple nodes and feeding the probe's connections to either a dedicated ArduPower collector unit per node (similar to 1) or by connecting probes of multiple nodes to a single collector unit. Traces then can be collected by the profiled node itself or by a dedicated node that all collector units are connected to.

B. ArduPower v2 Firmware

Microcontrollers like the Atmel ATmega2560, which is the processor used on the Arduino Mega platform, not only can interact with hardware like Hall-effect sensors, but also can be flashed with a custom firmware that enables processing of data. While ArduPower provides an Arduino-compatible shield and probes to plug into a computer's wiring harness, a complementary firmware is required to control the serial communication with a monitoring node, optimize data for efficient transport and enable features like automatic configuration.

Serial Protocol: A standard serial interface uses a frame length of 8 bits for data. The result of the analog-digital-conversion has a precision of 10-bits. For better efficiency in serial communication, the channel-combination-protocol was developed for ArduPower v1. This protocol features a synchronization bit and splits data of two probes to fit into three serial frames. This protocol leaves one bit unused in every three frames and cannot fill 4 bits of a frame for every uneven number of probes, leaving

$$\lfloor n/2 \rfloor + (n \mod 2) * 4$$

bits unused per n probes. By altering the serial connection to use 6 data bits, a 10 bit result can be split into two 5 bit frames and therefore each probe requires two serial frames. This change uses 100 % of all available data bits while having a 4 bit shorter overall length if the number of probes is odd.

C. Collector Program

While the hardware collects data on power consumption, calculation of the actual current, storing of values and analysis are handled externally by either the analyzed node itself or by a dedicated monitoring computer. To retrieve data, a collector program is provided, which handles receiving data from the ArduPower unit over a serial connection and deciphering the protocol. Also, the automatic configuration results are stored and used to determine power consumption.

All measurement values sent by the ArduPower unit are raw 10 bit ADC values, representing the output voltage of the ACS723 sensors. This value first has to be used to derive the actual sensor's output voltage, which then must be mapped to a current. This is done by solving

$$I = \frac{\text{type}}{2} * \left(\frac{\text{analogReadVal} - 5\,\text{V}}{1{,}024\,\text{V}} - 2.5 \right)$$

where `type` is the sensor's range (e.g., 20 A or 40 A) and `analogReadVal` is received from the ArduPower unit. Furthermore, the program enables starting and stopping of read cycles on the ArduPower unit, can request the current sensor configuration and handles synchronization errors.

V. Evaluation

Evaluation is done by comparing the results of our approach to similar works. Song et al. profiled the HPC Challenge benchmark set and can be used for cross-comparison of power draw behavior [15]. Comparing against the now deprecated ArduPower v1 platform shows improvements of the new device and allows to check the accuracy of the gained results.

For testing, a Nehalem based Intel machine was used. The system was equipped with two Xeon X5560 CPUs with a rated Thermal Design Power (TDP) of 95 W each. The TDP provides the maximum heat output of a chip in reference to the base clock frequency, therefore it is not the maximum electrical power draw of a chip, which can be higher. Power draw increases quadratically with frequency, therefore resulting in a much higher power draw than in the product specification when turbo boosting.

Data was collected from ArduPower v2 using a standalone collector script. However, a custom collector for the Diamond daemon is also available in the paper repository at [16]. It also includes ArduPower v2's firmware, schematics and more.

When interpreting the following results, a preliminary remark must be kept in mind, namely, the expected behavior in a dual socket system would be nearly identical power draw on both CPUs. As the boards feature two EPS connectors for power delivery to the CPU, the wattage on both should be the same. All experiments show an offset between both EPS lines, which leads to the assumption that power delivery to the CPUs is asymmetrical on both connectors. The difference between both inputs is a constant of 20 W, therefore it is assumed that this is due to the mainboard design.

A. High-Performance LINPACK

Aiming to rate the peak performance R_{peak}, the High-Performance LINPACK Benchmark (HPL) was developed. The benchmark in its core is a solver for dense linear equations leveraging LU factorization with partial pivoting, which then results in a number of 64-bit floating point operations that can be performed per second (FLOPS). The most common implementation of the LINPACK Benchmark was developed by Petitet et. al. and leverages MPI for scalability [17].

The LINPACK benchmark works in a loop with calculation and communication phases. While a computation phase applies load to the CPU, in the communication phase, performance data is shared over MPI, resulting in waiting times in different processes, due to non-parallel execution times.

In Figure 2, these phases can be seen as local minima. The total CPU power consumption peaks at close to 280 W. This

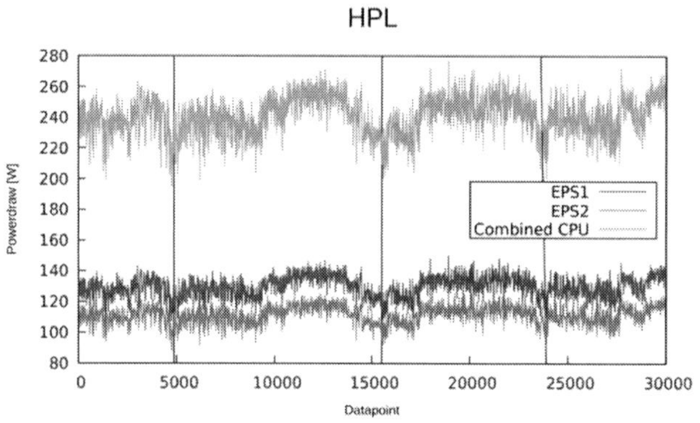

Figure 2. Excerpt of a HPL run measurement. The shown run time is 27.6 s with a total of 30,000 samples. Vertical lines mark the communication sections. Only every 15th sample is plotted for better visibility.

is much higher than the expected TDP of 190 W, but backed by the measurements of ArduPower v1, which also detects wattage peaks. The excess of 90 W therefore can be explained by the aforementioned Turbo Boost features, which overclock the CPU by 0.4 GHz over the base clock. Over the course of the run no thermal throttling could be observed, as this would be recognizable by valleys in the graph. In the future, information about application phases could be used to improve application performance, as well as energy efficiency. For instance, during computation phases, the network is typically idle and could be used for asynchronous I/O or other tasks.

B. MPI Fast Fourier Transform

Similar to the HPL, the MPI Fast Fourier Transform (FFT) benchmark is also contained in the HPC Challenge benchmark suite. It implements the fast Fourier transform operation over all MPI processes and can be used to identify scaling issues and MPI/communication-link limitations.

The MPI FFT's power draw graph is shown in Figure 3. The general course of power draw is a spike, which is a short computation or generation phase, followed by a plateau that represents MPI activity in communicating data. The measurements were done on a single node, which impacts the timings of these phases. As they are tightly linked to a single system's load, huge variation can be expected. The wattage in these phases is lower because of synchronization on MPI processes and memory copy operations that normally are not handled by the CPU. Again, information about computation and communication phases could help to use the system more efficiently. For instance, Turbo Boost could be selectively switched on for the short computation spikes.

C. High Performance Conjugate Gradients

While benchmarks such as the HPL test the performance of computation systems by producing huge arbitrary loads, the High Performance Conjugate Gradients benchmark proposed by Dongarra et. al. targets to model data access patterns as found in the real world by leveraging sparse matrix calculations and local Gauss-Seidel smoothing, besides other commonly used engineering methods [18]. The HPCG benchmark is known to consume much less power than the HPL, while maintaining a more stable draw [19]. In contrast to the HPL, this benchmark is memory bound, without the aim to calculate peak system performance. The results in Figure 4 are again confirmed with ArduPower v1, cross-validating the measurement.

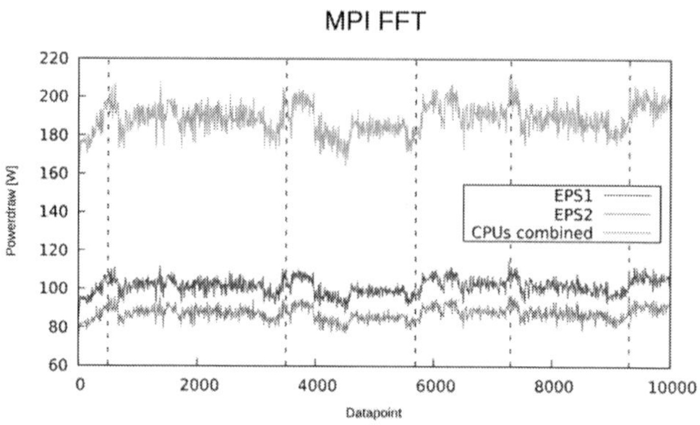

Figure 3. Power consumption of the MPI FFT test in the HPCC benchmark. Vertical lines signal peaks, which indicate computation/generation phases, followed by communication activity. The data is reduced to show every 10th sample.

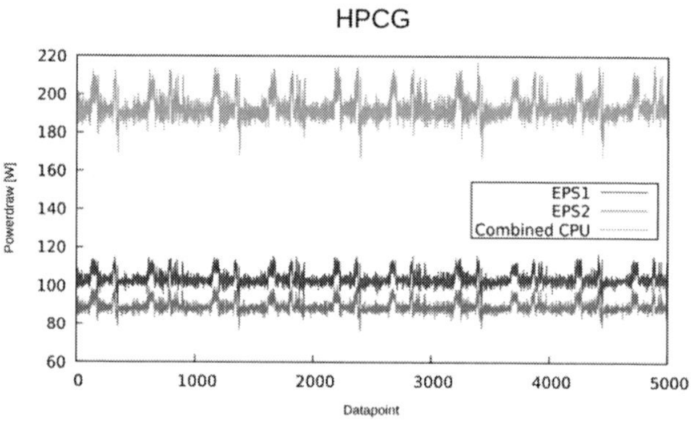

Figure 4. Excerpt of a HPCG power draw analysis. The shown duration is 4.6 s with a total of 5,000 samples. This plot shows every captured data sample.

D. ArduPower v2 Performance

Redesigning the ArduPower platform was done with the goals of increased usability, higher modularity for any given use case and increased performance. To increase throughput of captured data, a new protocol based on 6-bit serial communication was implemented. The sampling speed is mostly determined by the analog-digital-conversion. The new serial protocol on the other hand only increases performance marginally, as the effect only is noticeable with odd probe counts. The achievable sample rate with a varying number of connected probes is shown in Figure 5.

VI. Conclusion

ArduPower v2 is a modular and open design for an internal wattmeter that allows power measurements on a component level. We make use of an Arduino Mega 2560, as well as custom-built shield and software to enable high-frequency measurements using probes. In contrast to coarse-grained external solutions, this allows us to capture and correlate the power impact of individual application phases. This information can then be used to improve the energy efficiency of the overall software stack. ArduPower v2 is an evolutionary improvement of our previous design that delivers verified results while providing a richer feature set than its predecessor.

Since we follow an open approach, we make all necessary schematics, firmware, scripts, etc. available in a repository at

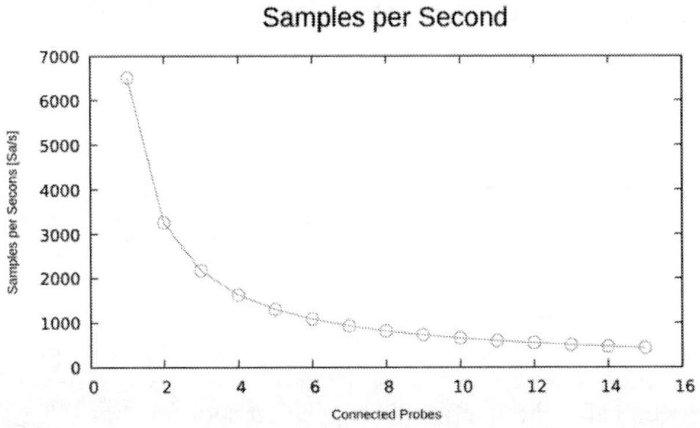

Figure 5. ArduPower v2 sample collection performance. The graph shows the achieved samples per second for each probe configuration. The rapid falloff is due to rising analog-digital-conversion impact on the cycle duration.

[16]. Interested parties can use the provided information to build their own version of ArduPower and use it for fine-grained power measurements. We welcome collaboration to improve both the hardware architecture itself, as well as its integration into the software stack. ArduPower v2 should be usable for computer clusters of any size and can be integrated without having to modify existing hardware. It can therefore provide a convenient building block for research on performance analysis and energy efficiency across many different fields.

VII. Future Work

Being built around modularity, the ArduPower platform can be extended to provide even more insights. In its current state, the focus is on internal power measurement, but a probe for mains electricity power consumption can be implemented to enable efficiency rating of power supply units and easier overall power consumption figures. Currently, there is no possibility to capture power draw from the mainboard to a component. Sophisticated adapters for memory slots, PCI-e or even M.2/U.2 slots, can be made to break down power consumption in even more detail. The open, modular design can also be used for analysis of different performance factors. Temperature probes tracking thermals and cooling capacity, anemometers for airflow analysis or even security mechanisms like intrusion detection can be developed and connected with via the simple five-pin interface. Besides analog values, as shown in the project, also digital communication, such as UART/USART or One-Wire, can be realized. Firmware modification could even allow SPI by providing a clock signal on the voltage return line.

Moreover, ArduPower could be used to provide additional functionality within existing software, such as SLURM. Due to its open design and modest costs, even clusters not equipped with vendor-specific solutions could be upgraded using ArduPower to enable fine-grained power measurements. In contrast to the first version of our approach, ArduPower v2 does not require cutting cables and can thus be integrated and removed easily. While ArduPower v2 still must be embedded into the actual system, its small form factor fits even into relatively small cases.

Due to the high resolution of ArduPower v2's captured measurements, it becomes possible to observe even small variations in power consumption, such as during waiting phases making use of spinlocks. Moreover, the data would make it possible to automatically determine even short application phases. This information could then be used to tune the underlying hardware appropriately.

References

[1] "Green 500 November 2019," Nov 2019, retrieved: April, 2020. [Online]. Available: https://www.top500.org/green500/lists/2019/11/

[2] D. N. S. S. Liyanage, G. V. M. P. A. Fernando, D. D. M. M. Arachchi, R. D. D. T. Karunathilaka, and A. S. Perera, "Utilizing Intel Advanced Vector Extensions for Monte Carlo Simulation based Value at Risk Computation," in International Conference on Computational Science, ICCS 2017, 12-14 June 2017, Zurich, Switzerland, ser. Procedia Computer Science, vol. 108. Elsevier, 2017, pp. 626–634. [Online]. Available: https://doi.org/10.1016/j.procs.2017.05.156

[3] "Slurm Workload Manager." [Online]. Available: https://slurm.schedmd.com/

[4] Diouri et al., "Assessing Power Monitoring Approaches for Energy and Power Analysis of Computers," Sustainable Computing: Informatics and Systems, vol. 4, no. 2, 2014, pp. 68–82.

[5] Intel, Hewlett-Packard, NEC, and Dell. IPMI Specification V2.0 Rev. 1.1. Retrieved: April, 2020. [Online]. Available: http://www.intel.com/content/www/us/en/servers/ipmi/ipmi-second-gen-interface-spec-v2-rev1-1.html (2013)

[6] D. Watts, P. Ainsworth, S. Aithal, B. Choi, and M. Rodriguez, "IBM Systems Director Active Energy Manager Installation and User's Guide," Oct 2019.

[7] Z. Electronic Systems, "4 Channel Power Meter LMG450," User Manual, 2009.

[8] J. H. Laros, P. Pokorny, and D. DeBonis, "PowerInsight - A commodity power measurement capability," in 2013 International Green Computing Conference Proceedings, June 2013, pp. 1–6.

[9] D. Bedard, M. Y. Lim, R. Fowler, and A. Porterfield, "PowerMon: Fine-grained and Integrated Power Monitoring for Commodity Computer Systems," in Proceedings of the IEEE SoutheastCon 2010 (SoutheastCon). IEEE, 2010, pp. 479–484.

[10] R. Ge et al., "PowerPack: Energy Profiling and Analysis of High-Performance Systems and Applications," IEEE Transactions on Parallel and Distributed Systems, vol. 21, no. 5, May 2010.

[11] D. Hackenberg et al., "HDEEM: High Definition Energy Efficiency Monitoring," in 2014 Energy Efficient Supercomputing Workshop, Nov 2014, pp. 1–10.

[12] S. J. Martin and M. Kappel, "Cray XC30 Power Monitoring and Management," in Cray User Group Conference Proceedings, 2014.

[13] M. Knobloch, M. Foszczynski, W. Homberg, D. Pleiter, and H. Böttiger, "Mapping fine-grained power measurements to HPC application runtime characteristics on IBM POWER7," Computer Science - Research and Development, vol. 29, no. 3, Aug 2014, pp. 211–219. [Online]. Available: https://doi.org/10.1007/s00450-013-0245-5

[14] M. F. Dolz, M. R. Heidari, M. Kuhn, T. Ludwig, and G. Fabregat, "ArduPower: A Low-Cost Wattmeter to Improve Energy Efficiency of HPC Applications," in Sixth International Green and Sustainable Computing Conference, IGSC 2015, Las Vegas, NV, USA, December 14-16, 2015. IEEE Computer Society, 2015, pp. 1–8. [Online]. Available: https://doi.org/10.1109/IGCC.2015.7393692

[15] S. Song, R. Ge, X. Feng, and K. W. Cameron, "Energy Profiling and Analysis of the HPC Challenge Benchmarks," Int. J. High Perform. Comput. Appl., vol. 23, no. 3, August 2009, pp. 265–276. [Online]. Available: http://dx.doi.org/10.1177/1094342009106193

[16] "ArduPower Github Repository." [Online]. Available: https://github.com/wr-hamburg/energy2020-ardupower

[17] A. Petitet, R. Whaley, J. Dongarra, and A. Cleary, "HPL – a Portable Implementation of the High-Performance Linpack Benchmark for Distributed-Memory Computers," 01 2008.

[18] M. A. Heroux and J. Dongarra, "Toward a New Metric for Ranking High Performance Computing Systems," 6 2013.

[19] "HPCG Performance Efficiency on VE at 5.99%," Nov 2019, retrieved: April, 2020. [Online]. Available: https://sx-aurora.github.io/posts/hpcg-tuning/

Large-Scale Co-Simulation of Power Grid and Communication Network Models with Software in the Loop

Eric MSP Veith[1], Jawad Kazmi[2], and Stephan Balduin[1]

[1] OFFIS e.V.
Oldenburg, Germany
Email: {veith,balduin}@offis.de

[2] Austrian Institute of Technology (AIT)
Vienna, Austria
Email: jawad.kazmi@ait.ac.at

***Abstract*—Power grids are transitioning from an infrastructure model based on reactive electronics towards a smart grid that features complex software stacks with intelligent, pro-active and decentralized control. As the power grid infrastructure becomes a platform for software, the need for a reliable roll-out of software updates on a large scale becomes evident. In order to validate resilient large-scale software roll-out protocols, corresponding test beds are needed, which mirror not only Information and Communication Technology (ICT) networks, but also include the actual software being deployed, and show the interaction between the power grid and the ICT network during the roll-out, and especially during roll-out failures. In this paper, we describe the design implementation of a large-scale co-simulation test bed that combines ICT and power grid simulators. We pay specific attention to the details of integrating containerized software in the simulation loop.**

Keywords*—*Co-Simulation; Smart Grid; Power Grid Information and Communication Technology; Software in the Loop; Linux Development

I. Introduction

The transition of the power grid to the smart grid is happening on a large scale. From the first introduction of the term *smart grid* [1], assets in the power grid have evolved into software platforms that feature a vast array of services. Transformers have become tools in asset management [2], while Multi Agent Systems (MAS) represent nodes in the power grid [3].

The numerous use and business cases enabled by this kind of infrastructure obviously require special attention to the software stack deployed on these devices. The life and, hence, innovation cycle in the power grid of 30–60 years that was dominating in the traditional power grid does not hold anymore. As the evolution of energy systems to Cyber-Physical Systems (CPS) based on ICT technologies has happened, so has, with increased complexity, risen the inherent risk of the overall system [4]: Power grids have become a target in terms of cyber security, as proven by the attacks on the Ukrainian power grid between 2015 and 2017 [5][6]. Specifically, software solutions based on Artificial Intelligence (AI) technologies have been regarded as major factors in technical debt, causing frequent updates to be made [7][8].

In a recent literature survey, we noted that the emerging smart grid yields numerous attack vectors, many stemming from the inclusion of ICT, AI technologies or tight market integration [9]. A major research gap exists in AI-based analysis of complex CPS, i.e., the combination of power grid and ICT. Specifically, the interaction of both components has hitherto seldom been discussed. On this basis, Adversarial Resilience Learning (ARL) offers an approach *based* on AI to explore any CPS without domain-specific knowledge and find weaknesses in its configuration [10][11]. This can very well be applied to software roll-out and update processes, too, provided a test bed for this exists. Software update roll-outs are, for a simulation testbed, a special case, as they require the actual software to be deployed within the simulation in order to assess the impact of the roll out.

To this end, we present a co-simulation approach that features power grid, ICT, and software-in-the-loop simulators. We will detail the specific development to facilitate a software-in-the-loop simulation on a large scale. The rest of this paper is structured as follows: Section II provides context for this work. We will detail possible, generalized models for our testbed in Section III. We then offer insights into the ICT co-simulation in Section IV, which accounts for a major portion of this paper. We discuss the overall development in Section V, and conclude with pointers to future work in Section VI.

II. Related Work

Simulators for specific domains exist for many years now, drawing from the standard rationale that, once the system and the interaction of its components become too complex to describe them in terms of formulæ and automatons, a simulation to assert assumptions is in order. For each individual domain, a sound selection of simulators exist, such as *pandapower* by Thurner *et al.* [12] and *SIMONA* by Kittl *et al.* [13] for power grids, or OMNeT++ by Varga *et al.* [14] for ICT simulations.

However, to witness effects of the two domains interacting with each other, none of the two is fully suited. Specifically when smart grid messaging is considered—which is crucial to optimization protocols, such as COHDA [15][16] or Winzent [17]—, this part of the simulation becomes crucial. Previous simulation environments for testing smart grid messaging have focused on other parts of the problem, such as using a Geospatial Information System (GIS) layer to model the feed-in of renewable energy sources [18]. Since a modern power grid has, essentially, become a CPS, ICT has become an integral part. Therefore, the interaction between these two complex domains has become paramount for our research.

The combination of two or more simulators from different domains is facilitated through *co-simulation*. A co-simulator provides an infrastructure to schedule, synchronize different simulators, and enable data exchange between model instances run by the different simulators. One solution is provided by the *mosaik* co-simulator [19]—the one, in fact, used to develop the test bed presented in this paper—; other approaches to co-simulation are employed, e.g., by *OpSim* [20], or *PTOLEMY II* [21]. A co-simulation of power grid and ICT has been described by different authors using different pieces of software [22, 23, 24], but without taking the question of software roll-out into account.

This paper introduces a smart grid software roll-out testbed, based on the idea discussed by Kintzler *et al.* [25]. It details the reasoning behind using a Software-In-the-Loop (SIL) approach—namely, that the software being rolled out itself is complex enough that an approximation through models is not feasible. If the subject to the experiment, i.e., the software, is abstracted away, the result of the roll-out protocol cannot be validated.

SIL co-simulation is not new. Pieper *et al.* [26] use the SIL technique to validate railway controllers; real-time SIL co-simulation in the smart grid for performance measurements is done by, e.g., Bian *et al.* [27]. The OMNeT++ simulation environment offers facilities for Hardware in the Loop (HIL) integration [28][29]. However, when the software itself is subject to change in a co-simulation/SIL scenario, an extension needs to be developed to allow the integration of changing, virtualized software containers. This research gap is addressed by our solution.

III. Modelling Power Grid and ICT

Figure 1 shows the data exchange schema of our co-simulation approach, including all software bridges that connect the simulation with the SIL part. The following sections will refer to the schema when locating individual pieces of software.

A. Power Grid Reference Model

To capture the complex dynamics of and possible effects caused by the large-scale roll-out of smart power devices, a realistic and complete model of the power system model is a necessity. The model needs to be detailed that could later be simulated along with the other related components. It is, therefore, important to choose a modeling and simulation tool that fulfills these requirements. There are many good power system modeling and simulation tools such as *pandapower* by Thurner *et al.* [12] and *SIMONA* by Kittl *et al.* [13] for power grids. After a survey and discussion, *DIgSILENT PowerFactory* was selected for the power system modeling as it meets the selection criteria better than the other available tools. It provides detailed and fine-grained modeling and simulation of many aspects of the power system. The model needs to be simulated in a co-simulation setting and allow to receive the set-point and measurements from and to the coupled sub-system, as shown in Figure 2.

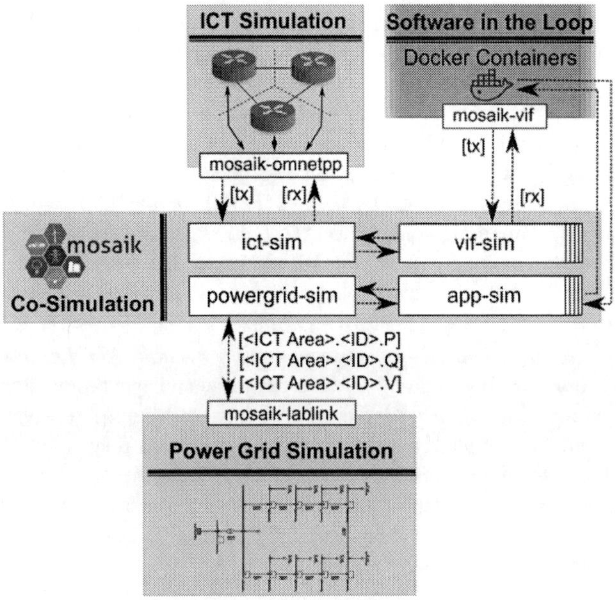

Figure 1. Data Exchange Schema of the Co-Simulation

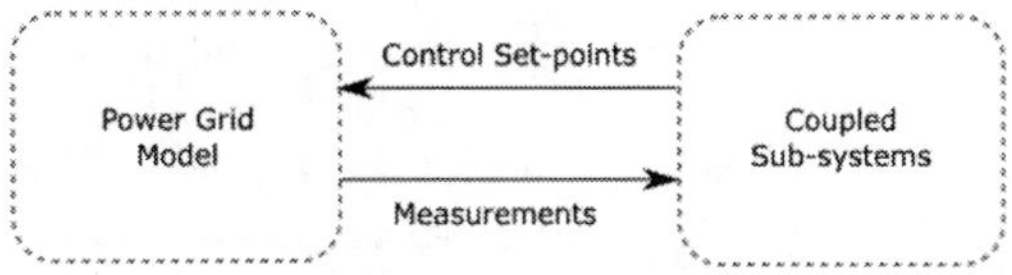

Figure 2. Power grid (co-)simulation design rationale

DIgSILENT PowerFactory is a sophisticated, highly specialized, flexible, and extendable platform for power system modeling and simulation. It supports fine-grained power system modeling and simulation through a combination of both graphical and scripting based methods for almost all the major areas of the power system, including generation, transmission, distribution, etc. There is a large library of models available that can be extended by writing custom components using the DIgSILENT Simulation Language (DSL). For a dynamic simulation of the power system, the tool provides many functionalities including load and power flow calculations, reliability and contingency analysis, and many more. The tool also supports Application Programming Interfaces (APIs) that can be used to communicate with other simulators. It further supports the automation using DIgSILENT Programming Language (DPL).

AIT Lablink is a multipurpose, highly efficient, and distributed middleware for coupling both hardware and software components in a co-simulation. It is used for coupling the individual components and thus makes the power grid simulation flexible and extendable. *AIT Lablink* provides interfaces, simulation control, and data exchange capabilities. By using it, it is possible to do either simulations or an emulation and it has been used extensively [30, 31, 32] for performing various validation and verification activities. A large set of hardware and software components is already supported by bridges that make extending the testbed very easy.

In the present setup, depicted in Figure 1, *AIT Lablink* provides a message bus that the participating components (software/hardware) can connect to through a bridge. This bridge facilitates the data exchange and simulation control, including the synchronization. The bridge and the participating component have a one-to-one correspondence. Two important such bridges are the *DIgSILENT PowerFactory* and *mosaik* bridge. There are some other *AIT Lablink* system components like *Synchronizer*, *Simulation Manager*, etc., that provide useful services, but are excluded here for brevity, as they are part of every setup created with *AIT Lablink*.

As the co-simulation is managed by *mosaik*, *AIT Lablink* coordinates with *mosaik* for data exchange and synchronization of the simulation. All the data exchange requests received from the coupled systems through *mosaik* are forwarded to the respective component (*DIgSILENT PowerFactory* in this case), while simulation synchronization requests are forwarded to *Synchronizer* that takes care of running the co-simulation components in sync.

B. *The Communications Infrastructure Model*

The ICT model serves to provide a number of realistic network areas to test the software roll-out scenarios. It is independent from the test bed software, i.e., it was developed in parallel as part of the test bed, but can be used on its own, e.g., without software in the loop. It features a number of subnets, with each subnet area designating a certain characteristic network environment, such as a well-built fibre channel network or a spotty wireless area. To this end, it models an Autonomous System (AS) with routers and intra-AS traffic/routing. These subnets have real IPv4 addresses assigned, as the ICT model needs to process actual Internet Protocol (IP) traffic generated by the existing software. The ICT infrastructure network is fully contained in the class C subnet

`10.64.0.0/10` .

Table I contains the relevant subnet specifications for the areas that are described in the following paragraphs.

The reason for choosing this particular kind of subnet is its rather remarkable subnet range and the fact that `10.64.0.0/10` is seldom used as an IPv4 address space. This way, the ICT model does not collide with existing private, class C IPv4 addresses, such as those assigned by Virtual Private Network (VPN) software. This leaves room for 8192 subnets with 254 hosts each in every defined network. The `/24`-subnet should be the only network size, regardless of how many hosts are contained in it. Routers always get the lowest IP addresses assigned, i.e., `.1`, `.2`, `.3`, etc., before the first hosts are added.

The test bed consists of 3 areas, which differ by their Quality of Service (QoS) parameters. We assume that most visible traffic we consider is either based on the Transmission Control Protocol (TCP) or employs similar mechanisms. This especially means that the protocol features a retransmission algorithm. Since packet loss can be caused either by a low-quality link or by network congestion, delay (denoted by d) is, for the purposes of this test bed, the most describing parameter of a link (besides its data rate).

TABLE I. ICT MODEL SUBNET SPECIFICATION

Network	10.64.0.0/10
Network Range	10.64.0.1 – 10.127.255.255
Dedicated Network	10.64.0.0/12
Shared Links Network	10.80.0.0/12
High-Impairment Network	10.96.0.0/12
Misc./Unallocated	10.112.0.0/12

The first area is the *Dedicated Network Area*. The underlying assumption is that of the best possible infrastructure, where a grid operator has deployed dedicated ICT cabling. Thus, the network is of high quality. This does not only create a realistic scenario, but also serves as the test case for the whole simulation infrastructure. The assumed nominal data rate is 1 GBit/s; the delay is modeled stochastically per packet as:

$$d \sim 10 + 50 \cdot f_{\lambda}(x, 1) \text{ [ms]} . \quad (1)$$

The function $f_{\lambda}(x, 1)$ denotes the drawing of a random number from an exponential distribution.

The second designated area is the *Shared Links Area*. Here, we assume that a grid operator uses the public infrastructure, such as Internet-facing connections. While we can assume that the necessary security precautions are taken (e.g., by deploying a VPN solution and generally encrypting traffic), other traffic interferes with the QoS of the update traffic we examine. I.e., we can assume that there are occasional packet drops due to congestion. As such, we model the delay as the drawing of a random number from a normal distribution:

$$d \sim \mathcal{N}(250; 20) \text{ [ms]} . \quad (2)$$

The area is well suited for variable-situation test cases. The link data rate is still good, being at 1 GBit/s nominally.

The extreme end of the spectrum is modeled by the *High-Impairment Area*. It features low-datarate links (configurable from 50 kBit/s up to 100 MBit/s with frequent congestion. This area also models the deployment of wireless connections, such as 4G/CDMA 450 or similar technologies. It is characteristic for an area where the development of the infrastructure was hindered by, e.g., existing building situations, harsh terrain, cost constraints, etc. As such, there are frequent packet drops and even connection drops. The delay is modeled as:

$$d \sim \mathcal{U}[100; \infty] \text{ [ms]} , \quad (3)$$

i.e., the drawing of a random number from a uniform distribution with the interval $[100; \infty]$ (inclusive). A delay of infinity means the link is broken.

IV. ICT AND SOFTWARE-IN-THE-LOOP DEVELOPMENT

A. *Data Exchange Flows*

All simulators appear in the system twice, as Figure 1 suggests. For each simulator process—like the ICT simulation, the power grid simulation, or each containerized SIL—there exists also a representation as an entity object in *mosaik*. This object is responsible for connection data exchange channels as well as communicating with the simulator processes. Overall, there are at least four simulator processes with corresponding entity objects.

The **ICT Simulation** is responsible to run the communication network simulation. Some nodes are also providing an interface to the co-simulation as a bridge between the ICT simulation and the SIL components. I.e., it also injects real IP packets from the containerized applications into the simulation environment and reads packets received from other simulated nodes and transfers them back to the software containers. In Figure 1, it is represented as the `ict-sim` object during a *mosaik* run.

The **Power Grid Simulation** is responsible for calculating load flows and line loads. It receives data through *mosaik* from the actual applications. For example, an application representing an intelligent substation would appear in the power grid simulation as substation; the substation software would receive readings from the power grid simulation and issue setpoints to it. This data flow is depicted in Figure 1 as an exchange between the SIL container, the `app-sim` entity objects, and the `powergrid-sim`.

The **Application Simulators** each represent one containerized piece of software. They are not simulators in the strict sense, but the SIL component. The simulator is responsible for starting and stopping the containers gracefully, and also for setting and collecting data coming from the co-simulation or going to another simulator. Each application container has its own application simulator and, hence, a corresponding `app-sim` entity object.

Application logic will dictate communication with other containers. For example, a distributed real power schedule optimization heuristic like *Winzent* works on MAS basis, and, therefore, requires communication with other applications. I.e., the application containers are the logical connection between ICT and the power grid simulation. For the roll-out scenario, it is not sensible to modify the application software to be part of the ICT simulation directly. Thus, we leave the applications in the container undisturbed, and deploy a virtual network interface to connect the applications to the ICT simulation. This virtual network interface, called *vif* for short, also has a corresponding `mosaik-vif` entity object in the *mosaik* process. Thus, in our scenario, there exist exactly as many `app-sim` objects as there are `vif-sim` entities.

The connection between the virtual interfaces and the corresponding nodes in the ICT simulation is done in *mosaik*. Any ICT-related simulator offers at least one model that represents the respective node. These models have exactly two attributes, `rx` ("receive") and `tx` ("transmit"). *Attribute* is a *mosaik* term that designates a data exchange interface for a simulator. Referring back to Figure 1, we see that each `vif-sim` has these two attributes. The `rx` attributes always receive data from *mosaik*, whereas `tx` attributes transmit data to *mosaik*. The ICT simulation has more than one `tx`/`rx` pair: one for every node in the simulation for which a corresponding application container exists.

```
someapp_vif_entity = \
    someapp_vif_simulator.vif()
someapp_ict_entity = next(
    x for x in ict_model.children
    if x.eid == \
        'SimulatedNetwork/SomeApp/app-0')

world.connect(someapp_vif_entity,
              someapp_ict_entity,
              ('tx', 'rx'))
world.connect(someapp_ict_entity,
              someapp_vif_entity,
              ('tx', 'rx'),
              time_shifted=True,
              initial_data={'tx': None})
```

Figure 3. Example of a connection between an entitity in the ICT simulation and the SIL container

An example of a connection in *mosaik* can be achieved as shown in Figure 3.

B. Virtual Network Adapter & Packet Injection

The code example in the previous section also shows how the hierarchical addressing for entities in simulators in *mosaik* is done. The `vif` entities here denote a SIL entity, i.e., a container with a unmodified piece of software. Each entity denotes two software instances: first, the virtual interface *vif* that exists in a container, and, second, the *vif-sim* that translates data between the container's networking stack and the *mosaik* co-simulation protocol. These two pieces of software must exists separately as to avoid timing issues. The startup behavior of the container and its software cannot be observed by the simulator; there exists a natural delay between launching the container and being actually able to integrate it in the simulation run, i.e., the containerized application sending and being able to receive packets. Since multiple containers will normally be started, there is a time gap between the first container's application being online and the last one being ready. As SIL implies no modification on the software, we cannot signal these applications to hold until the simulation is ready to be started; hence, each *vif* must transparently buffer all data until the *vif-sim* is launched by the co-simulator.

In general, the *vif* must act as if it was just a standard network device. For this reason, the Linux kernel's tun/tap device driver was chosen. It establishes a *tun device* that appears as `tun0` (or any higher index number) in the output of `ip address show`, can carry IPv4 and IPv6 addresses, and can be the subject of the default route. Moreover, the tun device needs no gateway address, i.e., `ip route add default dev tun0` without a `via` stanza is possible. This way, the tun device transparently receives all traffic from the application, which does not need to be changed; the kernel delivers all this traffic to a user space application, i.e., the *vif*. Injecting traffic is done the same way.

Since the userspace application needs to transmit this data to the co-simulator, a second, specific rule for the IP address

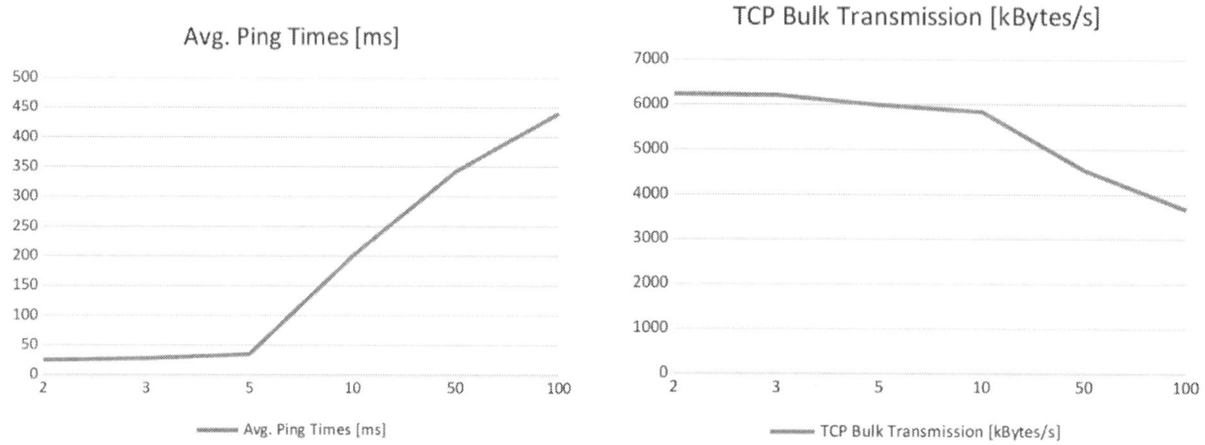

Figure 4. Experimental Measurements of Average Ping Times and Throughput

of the *mosaik* instance is added, so that traffic between the simulator and the *vif* still flows via the standard `eth0` device.

As the tun device now tunnels all regular traffic, the communication protocol between *vif* and *vif-sim* needs to be carefully chosen as to avoid race conditions. Tunneling TCP in TCP is discouraged, as two nested congestion control algorithms interfere with each other, creating cascading time lags that can stall the application, known as *TCP Meltdown* [33]. Since User Datagram Protocol (UDP) needs to be chosen, the external address of the container is not known to the co-simulator, which hinders the ICT simulation from injecting data first before any data is received from the container (and, thus, the container's address becomes known). We solve this be simply sending a burst of zero UDP 'hello' packets to announce the container.

Each byte of packet data received by the *vif* is immediately transmitted to the *vif-sim*, which takes care of assembling the packets. Assuming that the first transmission will contain the start of an IP packet and no intermediate packet data will be lost between container, *vif* and *vif-sim*, the *vif-sim* reads the packet length from the IP header field in order to assemble whole packets. These packets are then encoded in Base64 format so that they can be transmitted to *mosaik* via *mosaik's* Java-Script Object Notation (JSON) communication protocol.

The *vif-sim* as well as the *mosaik-OMNeT++* adapter are single-threaded, but use a cooperative, asynchronous I/O multitasking pattern to handle the communication flow. Under the assumption that these applications are I/O-heavy, but not computationally demanding, the single-threaded, multi-process paradigm where much time is spent in the kernel's I/O space suggests itself [34].

V. Discussion

As the general feasibility of co-simulation has already been established, we focused prominently on the ICT SIL simulations. For this, we have set up a co-simulation with a number of containers in which the *iPerf3* [35] application was running. We have deployed pairs of clients and servers so that an iPerf client can send and receive data from a dedicated iPerf server container. We used this set up to test both the average round-trip times (i.e., ping echo request/echo reply timings) and TCP bulk transfer speeds. All data was routed through the simulated ICT environment, so that the flow of data was as follows: *vif—vif-sim—mosaik—OMNeT++—mosaik—vim-sim'—vif'*. The simulated ICT environment does not impose additional artificial delays in its network model.

Figure 4 shows the behavior for both metrics given a rising number of nodes. Each data point represents a different number of nodes and the average over 100 repeated simulation runs. Delays rise sharply as the number of nodes rises, but not exponentially. With ping times in the area of 23 ms to 447 ms, we assume that applications that do not rely on real-time or, in general, low-latency communication can be accommodated by this SIL setup. However, the bulk throughput rate between 6102 kB/s to 3654 kB/s is far below a characteristic data rate normally achieved by standard Ethernet connections.

We have investigated the reason for the low data rate and have identified three major points. First, *mosaik* currently uses non-compressed JSON messages in a request-reply pattern for data exchange with out-of-process co-simulators. As both the *vif-sim* and the *mosaik-OMNeT++* adapter are written in C++, an additional network round-trip is introduced, even if the simulation runs locally. In addition, *mosaik's* single-threaded request-response communication pattern with its associated simulators means that dependent simulators expect a delay when other simulators are being stepped or queried for data.

Furthermore, *mosaik* has currently no facilities to allow simulators to signal the necessity to be stepped; simulator control is completely in the hands of *mosaik*. This means that *mosaik* must poll all *vif-sims* as often as possible, since the co-simulator has no other way of knowing when data is available from a SIL container. In contrast to the ICT simulation, data from applications arrives in a non-deterministic way. In general, we have observed delays in message processing stemming from the context switches between kernel space and user space that frequently occur as data from the containerized applications travel through several network stacks.

Moreover, we currently launch one *vim-sim* process per

container, as this is the easiest way from a software engineering organization perspective. However, this means a separate TCP connection per container, a new process, and a new data stream. We, therefore, plan to implement a multiplexing architecture in the *vim-sim* part in order to reduce the number of processes, and, hence, reduce task and context switches.

We believe that this approach offers great flexibility and ease in modelling ICT networks with SIL. As the development of *mosaik* is open source and already aimed at providing higher throughputs and lower delay in the communication with external simulators—e.g., a ZeroMQ implementation to replace the socket API already exists—, and the co-simulator is extended to allow for event-discrete, non-deterministic simulators as they exist in this scenario, we see an increase in the throughput in the near future.

VI. Conclusion & Future Work

In this paper, we have detailed requirements and issues encountered in a SIL co-simulation of software roll-outs in the power grid. We have shown how an interaction of ICT and the power grid can be simulated and how complete containerized software stacks can be embedded into this co-simulation.

In the future, we expect optimizations on implementation level, e.g., more efficient transports and serialization techniques, as well as implementing zero-copy primitives to reduce the number of copy operations and context switches. On a broader research perspective, we expect that abstracting parts of the system through surrogate models [36][37] will provide for a way to simulate large-scale roll-out procedures.

Acknowledgements

The presented work is conducted in the framework of the joint programming initiative ERA-Net Smart Grids Plus, with support from the European Union's Horizon 2020 research and innovation programme. On national level, the work was funded and supported by the Austrian Climate and Energy Fund (KLIEN, ref. 857570), by German BMWi (FKZ 0350012A), and by the Swedish Energy Agency (Project number 42794-1).

References

[1] S. F. Bush, *Smart Grid — Communication-enabled Intelligence for the Electric Power Grid*, 1st, ser. Wiley IEEE Series. Chichester, United Kingdom: John Wiley & Sons, 2014, ISBN: 978-1-119-97580-9.

[2] H. Ma, T. K. Saha, C. Ekanayake, and D. Martin, "Smart transformer for smart grid—intelligent framework and techniques for power transformer asset management", *IEEE Transactions on Smart Grid*, vol. 6, no. 2, pp. 1026–1034, Mar. 2015, ISSN: 1949-3061. DOI: 10.1109/TSG.2014.2384501.

[3] E. M. Veith, *Universal Smart Grid Agent for Distributed Power Generation Management*. Logos Verlag Berlin GmbH, 2017.

[4] O. Hanseth and C. Ciborra, *Risk, complexity and ICT*. Edward Elgar Publishing, 2007.

[5] R. M. Lee, M. J. Assante, and T. Conway, "Analysis of the cyber attack on the ukrainian power grid", E-ISAC, Washington, DC, USA, Tech. Rep., Mar. 2016.

[6] A. Prentice, *Ukrainian banks, electricity firm hit by fresh cyber attack*, Jun. 2017.

[7] D. Sculley *et al.*, "Machine Learning: The High Interest Credit Card of Technical Debt", *SE4ML: Software Engineering for Machine Learning (NIPS 2014 Workshop)*, pp. 1–9, 2014.

[8] D. Sculley *et al.*, "Hidden technical debt in machine learning systems", in *Advances in neural information processing systems*, 2015, pp. 2503–2511.

[9] E. Veith, L. Fischer, M. Tröschel, and A. Nieße, "Analyzing cyber-physical systems from the perspective of artificial intelligence", in *Proceedings of the 2019 International Conference on Artificial Intelligence, Robotics and Control*, ACM, 2019, pp. 85–95, ISBN: 978-1-4503-7671-6.

[10] L. Fischer, J.-M. Memmen, E. M. Veith, and M. Tröschel, "Adversarial resilience learning—towards systemic vulnerability analysis for large and complex systems", in *The Ninth International Conference on Smart Grids, Green Communications and IT Energy-aware Technologies (ENERGY 2019)*, vol. 9, 2019, pp. 24–32.

[11] E. M. Veith *et al.*, *Analyzing power grid, ict, and market without domain knowledge using distributed artificial intelligence*, 2020. arXiv: 2006.06074 [cs.CY].

[12] L. Thurner *et al.*, "Pandapower — an open-source python tool for convenient modeling, analysis, and optimization of electric power systems", *IEEE Transactions on Power Systems*, vol. 33, no. 6, pp. 6510–6521, Nov. 2018, ISSN: 0885-8950. DOI: 10.1109/TPWRS.2018.2829021.

[13] C. Kittl, J. Hiry, C. Wagner, C. Pfeiffer, C. Engels, and C. Rehtanz, "Large scale agent based simulation of distribution grid loading and its practical application", in *25th International Conference on Electricity Distribution (CIRED)*, Madrid, Spain: AIM, Jun. 2019, pp. 1–5.

[14] A. Varga and R. Hornig, "An overview of the OMNeT++ simulation environment", in *Proceedings of the 1st International Conference on Simulation Tools and Techniques for Communications, Networks and Systems & Workshops*, ser. Simutools '08, Marseille, France: ICST (Institute for Computer Sciences, Social-Informatics and Telecommunications Engineering), 2008, pp. 1–10, ISBN: 9789639799202.

[15] C. Hinrichs, S. Lehnhoff, and M. Sonnenschein, "A decentralized heuristic for multiple-choice combinatorial optimization problems", in *Operations Research Proceedings 2012*, Springer, 2014, pp. 297–302.

[16] A. Nieße, J. Bremer, and S. Lehnhoff, "On local minima in distributed energy scheduling.", in *FedCSIS Position Papers*, 2017, pp. 61–68.

[17] E. Veith, B. Steinbach, and J. Windeln, "A lightweight messaging protocol for smart grids", in *Proceedings of the Fifth International Conference on Emerging Network Intelligence (EMERGING 2013)*, Porto, Portugal, Sep. 2013, pp. 1–6.

[18] E. M. Veith, B. Steinbach, and J. Otten, "An open data-based discrete simulation environment for testing smart grid messaging", in *Proceedings of the Fourth International Conference on Smart Grids, Green Communications and IT Energy-aware Technologies (ENERGY 2014)*, ser. International Conference on Smart Grids, Green Communications and IT Energy-aware Technologies (ENERGY), Chamonix, France: International Academy, Research, and Industry Association, Apr. 2014, pp. 7–12.

[19] C. Steinbrink *et al.*, "Cpes testing with mosaik: Co-simulation planning, execution and analysis", *Applied Sciences*, vol. 9, no. 5, p. 923, 2019.

[20] S. R. Drauz, C. Spalthoff, M. Würtenberg, T. M. Kneikse, and M. Braun, "A modular approach for co-simulations of integrated multi-energy systems: Coupling multi-energy grids in existing environments of grid planning & operation tools", in *2018 Workshop on Modeling and Simulation of Cyber-Physical Energy Systems (MSCPES)*, IEEE, 2018, pp. 1–6.

[21] J. Eker *et al.*, "Taming heterogeneity-the ptolemy approach", *Proceedings of the IEEE*, vol. 91, no. 1, pp. 127–144, 2003.

[22] K. Hopkinson, X. Wang, R. Giovanini, J. Thorp, K. Birman, and D. Coury, "Epochs: A platform for agent-based electric power and communication simulation built from commercial off-the-shelf components", *IEEE Transactions on Power Systems*, vol. 21, no. 2, pp. 548–558, 2006.

[23] H. Lin, S. S. Veda, S. S. Shukla, L. Mili, and J. Thorp, "Geco: Global event-driven co-simulation framework for interconnected power system and communication network", *IEEE Transactions on Smart Grid*, vol. 3, no. 3, pp. 1444–1456, 2012.

[24] B. M. Kelley, P. Top, S. G. Smith, C. S. Woodward, and L. Min, "A federated simulation toolkit for electric power grid and communication network co-simulation", in *2015 Workshop on Modeling and Simulation of Cyber-Physical Energy Systems (MSCPES)*, IEEE, 2015, pp. 1–6.

[25] F. Kintzler *et al.*, "Large scale rollout of smart grid services", in *2018 Global Internet of Things Summit (GIoTS)*, IEEE, 2018, pp. 1–7.

[26] T. Pieper and R. Obermaisser, "Distributed co-simulation for software-in-the-loop testing of networked railway systems", in *2018 7th Mediterranean Conference on Embedded Computing (MECO)*, Jun. 2018, pp. 1–5. DOI: 10.1109/MECO.2018.8406023.

[27] D. Bian, M. Kuzlu, M. Pipattanasomporn, S. Rahman, and Y. Wu, "Real-time co-simulation platform using opal-rt and opnet for analyzing smart grid performance", in *2015 IEEE Power & Energy Society General Meeting*, IEEE, 2015, pp. 1–5.

[28] P. Wehner and D. Göhringer, "Internet of things simulation using omnet++ and hardware in the loop", in *Components and Services for IoT Platforms: Paving the Way for IoT Standards*, G. Keramidas, N. Voros, and M. Hübner, Eds. Cham: Springer International Publishing, 2017, pp. 77–87, ISBN: 978-3-319-42304-3. DOI: 10.1007/978-3-319-42304-3_4. [Online]. Available: https://doi.org/10.1007/978-3-319-42304-3_4.

[29] J. Kölsch, C. Heinz, S. Schumb, and C. Grimm, "Hardware-in-the-loop simulation for internet of things scenarios", in *2018 Workshop on Modeling and Simulation of Cyber-Physical Energy Systems (MSCPES)*, IEEE, 2018, pp. 1–6.

[30] I. Ahmad, J. H. Kazmi, M. Shahzad, P. Palensky, and W. Gawlik, "Co-simulation framework based on power system, ai and communication tools for evaluating smart grid applications", in *2015 IEEE Innovative Smart Grid Technologies - Asia (ISGT ASIA)*, 2015, pp. 1–6.

[31] J. H. Kazmi, A. Latif, I. Ahmad, P. Palensky, and W. Gawlik, "A flexible smart grid co-simulation environment for cyber-physical interdependence analysis", in *2016 Workshop on Modeling and Simulation of Cyber-Physical Energy Systems (MSCPES)*, 2016, pp. 1–6.

[32] M. Findrik, P. Smith, J. H. Kazmi, M. Faschang, and F. Kupzog, "Towards secure and resilient networked power distribution grids: Process and tool adoption", in *2016 IEEE International Conference on Smart Grid Communications (SmartGridComm)*, 2016, pp. 435–440.

[33] OpenVPN, *What is TCP meltdown?*, [Retrieved: 2020-07-08]. [Online]. Available: https://openvpn.net/faq/what-is-tcp-meltdown/.

[34] C. M. Kohlhoff, *Boost.asio*, [Retrieved: 2020-05-13]. [Online]. Available: https://www.boost.org/doc/libs/1_73_0/doc/html/boost_asio.html.

[35] J. Dugan *et al.*, *Iperf3*, [Retrieved: 2020-05-13]. [Online]. Available: https://iperf.fr.

[36] S. Balduin, "Surrogate models for composed simulation models in energy systems", *Energy Informatics*, vol. 1, no. 1, p. 30, 2018.

[37] S. Balduin, T. Westermann, and E. Puiutta, *Evaluating different machine learning techniques as surrogate for low voltage grids*, 2020. arXiv: 2006.12389 [eess.SP].

Intrusion Detection in Smart Grid Distribution Domain Using Deep Ruptures Detection

Sarah Chahine
Faculty of Engineering
University of Balamand
Balamand, El Koura, Lebanon
e-mail: sarah.chahine@std.balamand.edu.lb

Chafic Mokbel
Faculty of Engineering
University of Balamand
Balamand, El Koura, Lebanon
e-mail: chafic.mokbel@balamand.edu.lb

Abstract— **Smart grids brought huge added value to classical power grids in terms of advanced monitoring, metering control, trustworthiness and efficiency. This comes with some challenges, a major one being assuring the security of the grid against cyber-attacks. Obviously, such concerns are serious because of the impact and risks on electrical energy provisioning. To prevent and react to possible attacks, intrusion detection appears as a critical component. Previous literature work shows that an intrusion onto the grid translates into a small glitch that a phasor may help in identifying. In this work, we suggest to detect the glitches directly from electrical signals (current, voltage, frequency and power). We suggest using the detection of changes in the signals properties as an indicator of intrusion. To this end, classical approaches in ruptures detection have been experimented. A new approach based on deep Long Short-Term Memory (LSTM) filtering is proposed. The main focus of our work is on intrusions occurring in the distribution domain. In order to conduct experiments for the validation of the techniques, simulated data have been produced. The built simulator is also described in the paper. Benchmark results permit to confirm that our newly proposed deep nonlinear LSTM-based method is a viable solution to consider for intrusion detection for the distribution domain in a smart grid.**

Keywords-Cyber security; Smart Grid; Intrusion Detection; Ruptures Detection; Deep Filters; LSTM.

I. INTRODUCTION

The smart grid is the result of connecting the grid components using a communication network and extending the functionality through advanced monitoring, metering control and actuation devices. This permits to achieve better productivity, trustworthiness and efficiency [1]. Such achievement does not come without a major cost. In this particular case, concerns are raised about the security of the smart grid and the serious consequences of any cyber-attack [2]. With cyber facilities added, the grid does not only face power failures such as single line-to-ground, line-to-line, and two-phase-to-earth, but it also faces attacks on the security of the communication. If the network is compromised, an intruder can cause faults and more critical threats that endanger the power generation, transmission and distribution. A hacker getting control of the system may induce a big shut down, a change in the loads, a change in the pricing of the kilowatt and many more. The induced faults might be cascaded throughout the grid where a failure in one component can affect many others around it [3].

Cyber-attack on the smart grid can occur at different levels and in various domains. In this work, we are particularly interested in the attacks within the distribution domain. Jamei et al. [4] propose to build resilient Cyber-Physical Systems (CPS) by using tools that monitor and analyze data collected from Micro-Phasor Measurement Units (μPMU). The authors show that an intrusion onto the grid translates into a small glitch that a phasor may help in identifying.

In the present paper, we assume that an intrusion leads to glitches in the electrical signals. Thus, the detection of glitches alerts about possible intrusion. We suggest to collect normal electrical signals (current, voltage, frequency and power) from the distribution domain of a smart grid and to apply ruptures detection algorithms in order to localize glitches and, thereby, possible intrusions.

Detecting changes in stochastic signals has been a major domain of research in the past few decades [5]-[8]. Applications exist in several sectors. In order to detect an abrupt change in signal characteristics, two components are needed: i) a model providing a cost function and ii) a search strategy. The model generally compares the properties of local parts of the signals to background properties of the same signals. Parametric and statistical models or statistical hypothesis testing may be used. Once the model is identified and trained, a search strategy is needed in order to select and slide the local and background windows of the observed signals across time. In this paper, we use the ruptures open source to detect the glitches [8]. We suggest using two deep neuronal nonlinear predictive filters in order to estimate the costs of an abrupt change, i.e., the presence of glitches. An interface between our nonlinear filters and the ruptures is also being developed in order to apply the same search strategies.

The paper is organized as follows. Section II provides a short description of smart grids and cyber security. Section III describes the approaches used to perform the detection of intruders. Section IV presents the simulator that has been specifically developed to obtain data for the experiments. The experiments conducted in order to validate the approach and benchmark the different technologies are detailed in Section V. Finally, the paper ends with the major conclusions and several perspectives.

II. SMART GRIDS AND CYBER SECURITY

Kim et al. [9] define the smart grid as an electrical grid that couples the power system with an Information Technology (IT) system. This offers several advantages in terms of advanced monitoring, control and efficiency, but introduces new risks, a major one being the threats related to security.

A. *Smart Grids*

National Institute of Standards and Technology (NIST) identifies seven domains in a smart grid [10]: Generation, Transmission, Distribution, Customer, Operations, Markets, and Service Provider. The electrical power is generated at different generation stations in the generation domain. The transmission domain is where the energy produced is being transmitted to the consumers. The distribution domain is where the test feeder is situated, and it is the domain the customers are connected to. It is the place where the high voltages are lowered and regulated for common use. The customer domain is where the energy is finally consumed. Customers are the end users. These four domains: generation, transmission, distribution and customer, form the physical system of a grid. Electrical power lines connect the different components of these four domains. In this work, we are particularly interested in the distribution domain.

Besides the physical system, three other domains exist in a smart grid. Service providers are just like any type of service providers in a different sector. To illustrate this, one can compare them to the Internet service providers where they are the direct contact with the customers. The operations domain is where the controller of the grid resides. The controller receives measurement and other monitoring information about the grid through the communication network. Based on the received information, the controller takes informed decision and sends commands to the different controlled units in the grid. Finally, the markets domain is where the marketing issues related to power production and consumption are treated.

Secure network communication links interconnect all the components of the seven domains. This network allows devices, systems or programs to exchange necessary information and interact for executing advanced applications within the smart grid.

B. *Cyber Security*

The introduction of the cyber system to the smart grid did not just bring technological advancements, it also introduced new and critical problems. Now, the grid not only faces power failures, such a single line-to-ground, line-to-line, and two-phase-to-earth, but it also faces attacks on the security of the communication. This can let a hacker in, and that can introduce faults and more critical threats that endanger the power generation, transmission and distribution. A hacker getting control of the system may induce a big shut down, a change in the loads, a change in the pricing of the kilowatt and many more. The induced faults might be cascaded throughout the grid where a failure in one component can affect many others around it [3]. Generally, the concerns are about two major classes: power grid safety and data safety. Actually, a cyber-attack can occur on different nodes in any part of the grid and in any domain. It can be a distributed denial of service, false injection of data, gaining access and control over the system, tapping the system and eavesdropping on the data passing, spreading a malware and many more. In this work, the objective is to build an intrusion detection system to cope with attacks that might occur in the distribution domain in order to assure better power grid safety.

C. *Intrusion Detection*

The methodologies of intrusion detection are categorized as "Anomaly-Based Detection" (AD), "Signature-Based Detection" (SD), and "Stateful Protocol Analysis" (SPA) [12]. A "Signature-Based Detection" is a detection where the attack is known and, thus, the method to solve it is also known. It can be described as Knowledge-Based detection. The "Anomaly-Based Detection" identifies deviations from the expected behavior. The normal behavior is picked up from studying the background data for a while. It can be described as Behavior-based Detection. The "Stateful Protocol Analysis" may look like the "Anomaly-based Detection", however, it is based on knowing and tracing the protocol. SPA is also known as Specification-based detection.

Several studies have been conducted to build an Intrusion Detection System (IDS) for smart grids. Sedjelmaci and Senouci studied a combination of both distributed and centralized IDS with a focus on attacks such as Denial of Service (DoS), faulty data injection and resource injection [13]. They focused on the use of machine learning along with rule-based detection. They discovered that using rule-based alone consumed more energy, while the combination of machine learning and rule-based detection led to lowering the use of energy needed in detection. In addition, the detection turned out to be improved when combining both. Yu et al. presented an anomaly-based and watermarking-based IDS to counter false data injection [14]. Using watermarking, they insert hidden data to be able to verify the authenticity of the exchanged information and to detect any malicious injection.

Jamei et al. used microphasors to detect intruders [4]. Phasors are usually used in the transmission domain and much less in the distribution domain. The authors introduced microphasors in the distribution domain for cyber-attack detection. They compared the results using those microphasors to those from a distributed SCADA system.

Ozay et al. [15] use machine learning algorithms to detect complications in the smart grid system. These algorithms are also used to detect attacks and be able to differentiate faults from attacks. The fewer the False Positives and False Negatives, the higher the accuracy of the machine learning algorithm. The lack of data to train these machines presents a serious challenge. The use of machine learning algorithms is found extensively in a lot of researches to monitor and control systems [15]. It is also used now to detect attacks and be able to differentiate faults from attacks.

In the present work, we aim at building an anomaly-based IDS in the distribution domain using deep predicting filters that detect changes in measured electrical signals.

III. Intrusion Detection By Detecting Ruptures

As mentioned above, the proposed approach to intrusion detection relies on ruptures detection in electrical signals. Classical ruptures detection approaches are studied and experimented. We suggest performing rupture detection using a deep nonlinear filter.

The key idea in our intrusion detection is to catch a change in the electrical signals' properties, i.e., one or more of the measurements of voltage, current, frequency and power as a function of time.

Ruptures can be detected when changes in signal properties are observed. In order to achieve such detection, several approaches can be used. One approach consists in building a model representing the signals' background characteristics and to decide if such a model fits with each short window of the signals [7]. Another approach compares statistical properties or performs statistical tests in order to verify if the signals in different windows are the results of a unique process [8].

In this paper, we adopt the same formulation as in [8]. Let $\underline{y} = \{\underline{y}_1, \ldots, \underline{y}_T\}$ be a Multivariate non-stationary random process where $y_t \in \mathbb{R}^d$. It is supposed that $\underline{y}$ is piecewise stationary, i.e., there exist K unknown instants of ruptures $t_1^*, \ldots, t_K^*$ where some characteristics of $\underline{y}$ change. In order to determine K as well as the instants of ruptures, a criteria function is defined as:

$$V = \sum_{k=1}^{K-1} c\left(\underline{y}_{t_k}, \ldots, \underline{y}_{t_{k+1}}\right) \tag{1}$$

In (1), $c(.)$ is a cost function which measures goodness-of-fit of the signal segment to a specific model, as defined in [8]. If K is unknown, the cost is compared to a threshold in order to decide on the ruptures. The model represents the background information while local signals are to be tested against the background model in order to compute the cost. Once the model-cost function is defined, a search algorithm must be adopted. This algorithm defines how the local and background windows are set and used to explore the signals.

The models-cost functions are categorized into parametric and non-parametric [8]. Some parametric cost functions are: Maximum Likelihood Estimation, Multiple Linear Model, and Mahalanobis-type Metric. Some non-parametric cost functions are: Non-Parametric Maximum Likelihood Estimation, Rank-Based Detection, and Kernel-Based Detection.

In [8], search methods are classified into three different categories:

- Window-Sliding
- Binary Segmentation
- Bottom-Up Segmentation

Binary segmentation is "greedy sequential algorithm" [8]. The Bottom-up segmentation of a signal is used to perform fast segmentation of a signal. It works just the opposite of Binary segmentation. It starts first with multiple points of change and then decreases them by taking out the less important ones (the ones with least inconsistency) until there remains only the one that actually represents the correct number of changes. This is done by splitting the signal into various small sub-signals and then merging these parts sequentially until there remains the number of change points only. The window-sliding method computes the difference between two adjacent windows. The discrepancy can be described by the function below:

$$d(y_{a..t}, y_{t..b}) = c(y_{a..b}) - c(y_{a..t}) - c(y_{t..b}) \quad [8] \tag{2}$$

where $1 \le a < t < b \le T$

This function identifies how this approximation method measures the difference between one window and the one just after it. The distance is higher if the cost of the concatenation of the two adjacent windows is higher than the sum of the costs relative to each window taken separately.

For identifying the number of change points, certain constraints are taken based on whether the points of change are known or not. The accuracy of any detection method is the ability to estimate correctly the place of change points.

A. Classical Models

In the present work, the following models have been experimented for cost calculation:

- Autoregressive (AR)
- Least Absolute Deviation
- Least Squared Deviation
- Linear Model Change

An autoregressive model of order p computes an estimate of the present sample of a signal as a linear combination of the p previous samples. In case of scalar signals, this can be written as:

$$\tilde{y}_t = \sum_{i=1}^{p} a_i y_{t-i} \tag{3}$$

The autoregressive combination coefficients are determined in a way to minimize the mean square prediction error.

$$\underline{\hat{a}} = \underset{\underline{a}}{\operatorname{argmin}}\, Err(y_i \ldots y_f) = \underset{\underline{a}}{\operatorname{argmin}} \sum_{t=i+p}^{f} \|y_t - \tilde{y}_t\|^2 \tag{4}$$

The linear model change minimizes the mean square prediction error. Let $0 < t_1 < t_2 < \ldots < n$ be the unknown points of change. The linear regression model is described as:

$$y_t = \underline{z}_t^T \underline{\delta}_j + \varepsilon_t,\ \forall\, t = t_j, \ldots, t_{j-1} - 1 \text{ for } j > 1 \tag{5}$$

where y_t is the observed dependent variable, $\underline{z}_t$ is the covariate vector and $\underline{\delta}_t$ is the prediction coefficients vector.

The cost function over an interval I is the minimum mean square of the prediction error ε_t.

$$c\left(\underline{y}\right) = \sum_{t \in I} \|\varepsilon_t\|^2 = \sum_{t \in I} \|y_t - \underline{z}_t^T \underline{\delta}_j\|^2 \tag{6}$$

B. Deep Prediction Model

In addition to the previous models, two deep learning nonlinear models were also introduced with a customized cost function [8]. These are the Long Short-Term Memory

(LSTM) and the Multivariate machine learning. These machines have been used as nonlinear predictors of the current values of the electrical signals. It is assumed that, if the machines are well trained and if the signals do not change their properties, i.e., no glitch is present, then the prediction error shall be relatively small. In contrast, a large prediction error shall be observed when a glitch is present.

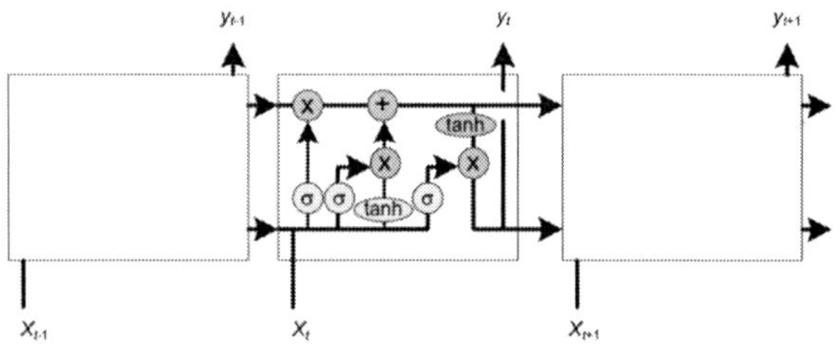

Figure 1. LSTM predictive filter (Reproduced from [16]).

The LSTM is a recurrent neural network. Its architecture is shown in Figure 1. Its name shows that it keeps information for a short term from past inputs. Because of its continuous learning process capability, it increases its resistance to noise and overcomes technical problems that might arise. The two practical complications overcome by LSTM are exploding and vanishing gradients, both linked to how the network is trained.

Multivariate predictive filter is a recurrent neural network as well [17]. Just like the LSTM, the Multivariate is resistant to noise and has a learning capability to be able to detect changes such as abrupt changes in the system. Figure 2 shows the relation of the different layers to the output layer. The purpose of this bidirectional neural network is to train the system both forward and backward while having both the forward and backward layers connected to the output.

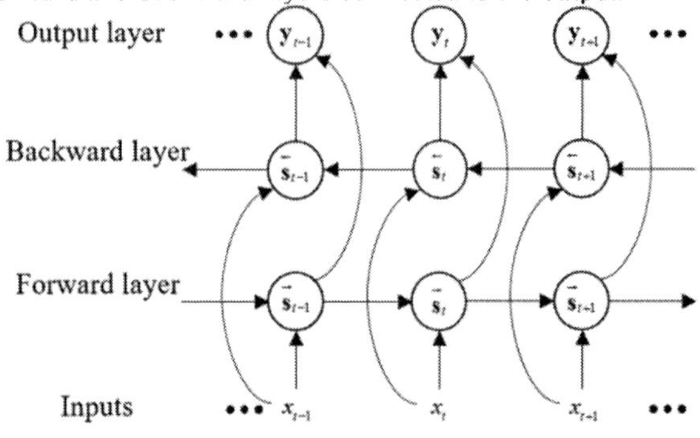

Figure 2. Multivariate prediction filter (Reproduced from [17]).

The LSTM and Multivariate are best used since they can take Multivariate signals, i.e., multiple electrical measurements. Convolutional layers may be added at their input in order to make the filter deeper. The predicted signal at their output serves to compute a cost function that is the mean square of the prediction error.

$$c(\underline{y}) = \sum_{t \in I} \|\varepsilon_t\|^2 = \sum_{t \in I} \left\| \underline{y}_t - \tilde{\underline{y}}_t \right\|^2 \quad (7)$$

where $\tilde{\underline{y}}_t$ is the predicted signal by the neural filter.

The mean square prediction error cost function is being used as for non-neural algorithms in order to detect the presence of ruptures in the electrical signals, which identifies possible intrusion.

IV. SIMULATOR

A simulator has been developed in order to generate the data that served in our experimentations. We focus on intrusions occurring in the distribution domain. However, a glitch resulting from an intrusion will be carried back to the main center where appropriate action shall be taken. Thus, the simulation must cover the distribution grid, the transmission grid and the control center. This helps narrow down the section where the problem is occurring. For this reason, the simulator system used here considers the three domains: transmission, distribution, and network.

The simulation system is made up from open-source components: Hierarchical Engine for Large-scale Infrastructure Co-Simulation (HELICS), Network Simulator – 3 (NS-3), GridDyn, and Gridlab-d, in addition to MATLAB. The HELICS acts as the main connector between all the different parts [18]. The NS-3 is used as the network of the system that connects the various parts [19]. The GridDyn [20] is used for creating and simulating the transmission grid. The Gridlab-d [21] and the MATLAB [22] are used for the creation and the simulation of the distribution grid.

HELICS is used to combine GridDyn (Transmission Grid) with NS-3 (Communication Network) along with MATLAB (Distribution Grid with static changes) or Gridlab-d (Distribution Grid with dynamic changes). To set up the system, a server and a client need to be created. The server collects and stores all the data. The client side is where all the action will be taken. The attack is taken up on the distribution grid and, for that reason, the system is created in a way to focus and grab data from that point.

After the setup of the system is completed, several tests are run where a simulation is created as if the hacker is going into the system. The attack creates a glitch for a fraction of time. Attacks in various time frames, phases, and grid states are simulated. A breaker switches on and off for a fraction of time to recreate the same glitch. Data is collected for the current, voltage, power and frequency. An XML file is generated for each test and keeps track of each experiment's context and conditions. For Gridlab-d, the IEEE 13-bus test feeder is used.

Jamei et al. [4] show how the glitch occurs and it was replicated with the setup provided.

Figure 3. Power Variation with Simulated Glitch.

Figure 3 shows a sample of the simulation where the power signal is generated with a glitch in it. The x-axis represents the time axis while the y-axis shows the power value for phases A, B and C of the power grid.

V. Experiments and Results

Using the simulator, forty different cases have been generated. For each simulation, the instance of the glitch is known. The electrical signals collected are voltage, current, frequency and power. A threshold value is specified on the cost in order to decide on a possible rupture. Ruptures detection algorithms may yield several consecutive detection points. A range of 500 points around the detected point is taken as the same point in order to smooth the detection results. Two types of errors can occur: a detection of rupture when it does not exist (False Acceptance) and a non detection of a rupture (False Rejection). False Acceptance (FA) and False Rejection (FR) rates are calculated based on the number of points detected and their location within the accepted range and based on the known reference points.

A. *Classical Ruptures Detection*

The forty simulations are fitted into five different models which are AR Order 4, Gaussian Process Change, Least Absolute Deviation, Least Squared Deviation and Linear Model Change. We have experimented several thresholds to which the cost functions are compared. In this paper, we report identical results on two different threshold values 5 and 0. The Gaussian Process Change was not taken into consideration when checking the FA and FR since most of the results did not give any break points, although they existed.

TABLE I. FA AND FR RATES FOR CLASSICAL RUPTURES DETECTION

		FA (%)	FR (%)
AR Order 4	Current	25557.30	0
	Power	20377.67	0
	Voltage	19819.06	0
Least Absolute Deviation	Current	8063.17	290.36
	Power	8861.98	0
	Voltage	11433.17	0
Least Squared Deviation	Current	6851.54	0
	Power	7695.77	0
	Voltage	8374.67	0
Linear Model Change	Current	9195.32	37.50
	Power	16865.10	0
	Voltage	21716.36	22.50

Table I shows the results of FA and FR for the different cost functions. For all the different cost functions, the number of falsely accepted points is high, which makes it far from acceptable for the real application. For the LSTM and Multivariate, the check gets updated every 200-point range. The error is calculated based on cost function the square root of the square sum. Below is the pseudo-code of the error() function.

```
def error(self, start, end):
  if end - start < self.min_size:
    raise NotEnoughPoints
  max = 0.0
  b = start
  s = 200
  while b <= end-s:
    sub=self.signal[b:b+s]
    y=np.sqrt(sum(np.square(sub)))/s
    if y[0] > max:
      max = y[0]
    b = b+s
  self.win_num = self.win_num+1
  return max

def sum_of_costs(self, bkps):
  epsilon=0.01
  for start, end in pairwise([0] + bkps):
    value=self.error(start, end)
    if value < epsilon:
      return value
  soc = max(self.error(start, end)
        for start, end in pairwise([0] + bkps))
  return soc
```

In the error function above, we define a start point and a window size. The error is calculated based on the square root of the square sum of this window. It is then compared with the threshold. If it surpasses the threshold, then this point will be considered as a glitch point.

TABLE II. LSTM PREDICTION ERROR SIGNALS USED AS INPUT OF RUPTURES DETECTION ALGORITHMS.

		FA (%)	FR (%)
AR Order 4	Current	263.11	3.40
	Power	189.23	5.46
	Voltage	190.82	6.98
Least Absolute Deviation	Current	234.61	1.03
	Power	297.20	3.05
	Voltage	256.15	0.31
Least	Current	391.13	0.67

Squared Deviation	Power	285.20	4.83
	Voltage	455.57	0.67
Custom Cost	Current	430.15	3.27
	Power	268.57	3.76
	Voltage	414.78	3.27
Normalized Custom Cost	Current	388.79	1.74
	Power	244.02	4.98
	Voltage	438.90	1.71

TABLE III. MULTIVARIATE PREDICTION ERROR SIGNALS USED AS INPUT OF RUPTURES DETECTION ALGORITHMS.

		FA(%)	FR(%)
AR Order 4	Current	234.17	4.74
	Power	165.72	6.23
	Voltage	267.07	5.50
Least Absolute Deviation	Current	217.39	4.07
	Power	263.01	4.12
	Voltage	328.85	4.12
Least Squared Deviation	Current	344.23	3.72
	Power	278.18	4.48
	Voltage	344.11	5.15
Custom Cost	Current	397.03	5.70
	Power	231.03	4.74
	Voltage	420.12	2.24
Normalized Custom Cost	Current	281.18	2.38
	Power	247.42	4.48
	Voltage	374.49	4.07

B. Deep Nonlinear Ruptures Detection

The LSTM and Multivariate predicted signals are being compared to the real signals and error signals are generated. These signals are fitted into different models which are AR Order 4, Gaussian Process Change, Least Absolute Deviation, Least Squared Deviation, Custom Cost (Square root of the square sum), and Normalized Custom Cost (Normalize the square root of the square sum). The rupture detection threshold is set to 0. The Gaussian Process Change in most of the cases does not deviate into giving a result and detecting any breakpoints. This makes it get ruled out when calculating the FA and FR afterwards.

The same experiments were conducted for Multivariate prediction. The Linear Model Change did not deviate at all and did not give results. The Gaussian Process Change reacted the same way as it did when it was applied to LSTM. For that reason, both the Linear Model Change and the Gaussian Process Change got removed when calculating the FA and FR later.

Tables 2 and 3 present the results in terms of FA and FR rates when the LSTM and Multivariate methods are first applied and then the generated prediction error signal is entered into the various cost functions, i.e., detecting changes of properties in the prediction error signal. Significant improvements are observed. However, the high FA makes it unacceptable as a precision for the electrical sector.

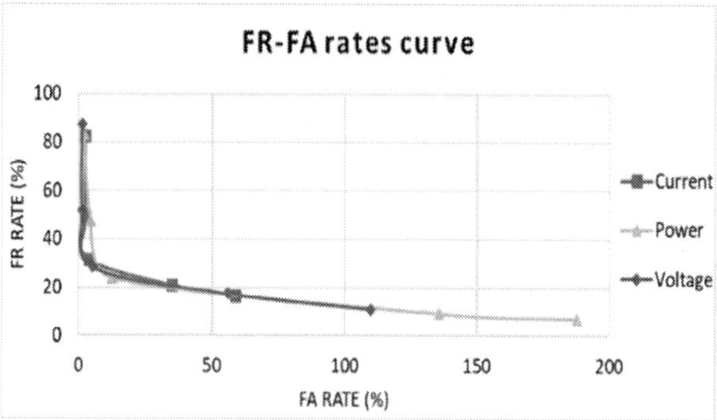

Figure 4. FR function of FA rates for LSTM with normalized cost.

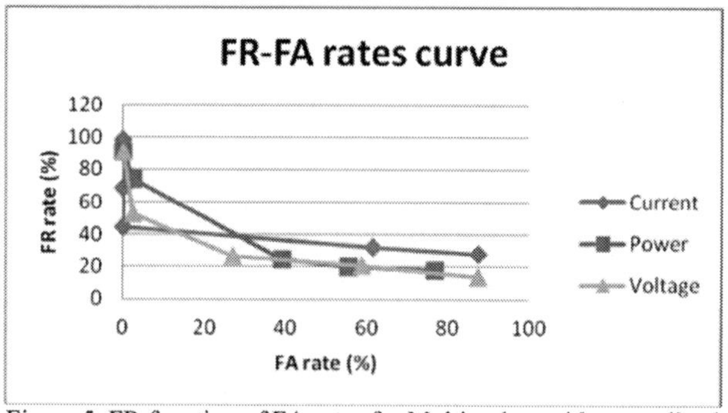

Figure 5. FR function of FA rates for Multivariate with normalized cost.

The prediction error signals obtained at the output of the deep nonlinear filters are now used directly to compute the cost. In this case, the cost becomes the mean square over the signal window (7). FA and FR rates are computed for different thresholds $\{0.01, 0.05, 0.07, 0.1, 0.2\}$.

Figure 4 presents the results of the LSTM method. The number of detected points is optimal when threshold is 0.07. A lower number of points is detected as the threshold increases over 0.07. A higher number of breakpoints are detected as the threshold decreases. Both FA and FR are low. The FR is the lowest which means that the system will rarely miss out any of the attackers getting into the system. This combination makes it the optimal thus far.

The same procedure is applied for the Multivariate method. The thresholds used are 0.01, 0.05, 0.07, 0.1, and 0.2. Figure 5 presents the data of the FA and FR rates obtained. The number of detected points is optimal when epsilon is 0.07. A lower number of points is detected as the threshold increases over 0.07. A higher number of breakpoints is detected as the threshold decreases. However, the performance is lower than for LSTM when the threshold=0.07.

By comparing all the above results, it can be concluded that the optimal method applied is when applying the LSTM, considering the cost as the mean square prediction error, and taking the threshold epsilon as 0.07. It gives the optimal result with a low number of FA. It also yields the minimal FR. This shows that the newly introduced modified system gives the most accurate results.

VI. CONCLUSIONS AND PERSPECTIVES

In this paper, we proposed and studied an Anomaly-Based IDS in the distribution domain that uses deep predicting filters applied on electrical signals to detect ruptures. We assumed that an intrusion in the distribution domain yields a glitch in the electrical signals that might be identified when applying rupture detection techniques.

In order to perform the study, a simulator has been developed and a set of forty signals have been generated representing a variety of grids profiles and intrusions. The simulator covers the three domains: transmission, distribution and networking.

Classical ruptures detection algorithms have been first applied on the signals and provided poor performances. A new deep nonlinear rupture detection technique has been thus proposed. Two deep prediction filters have been developed, an LSTM and a Multivariate recurrent network. Satisfactory results were obtained when mean square prediction error was considered as the cost function. LSTM seems to provide the best performance.

As a perspective, we plan to study a combination of the experimented detectors.

REFERENCES

[1] K. Wang et al., "Wireless Big Data Computing in Smart Grid," IEEE Wireless Communication, vol. 24, issue 2, pp. 58-64, April 2017, doi:10.1109/MWC.2017.1600256WC.

[2] M. B. Line, I. A. Tondel, and M. G. Jaatun, "Cyber Security Challenges in Smart Grids," 2nd IEEE PES Int. Conf. and Exhibition on Innovative Smart Grid Technologies, Dec. 2011, pp. 1-8, doi: 10.1109/ISGTEurope.2011.6162695

[3] D. Liu, X. Zhang, and C. K. Tse, "A stochastic model for cascading failures in smart grid under cyber attack," IEEE 3rd International Future Energy Electronics Conference, June 2017, pp. 783-788, doi: 10.1109/IFEEC.2017.7992139.

[4] M. Jamei et al., "Micro synchrophasor-based intrusion detection in automated distribution systems: Toward critical infrastructure security," IEEE Internet Computing, vol. 20, issue 5, Sep-Oct 2016, pp. 18-27, doi: 10.1109/MIC.2016.102.

[5] M. Basseville and A. Benveniste (Eds), "Detection of Abrupt Changes in Signals and Dynamical Systems," Lecture Notes in in Control and Information Sciences, LNCIS-77, Springer Verlag Berlin, Dec. 1985.

[6] R. Andre-Obrecht, "A New Statistical Approach for the Autoatic Segmentation of Continuous Speech Signals," IEEE Trans. On Acoustics, Speech amd Signal Processing, vol. 36, issue 1, Jan. 1988, pp. 29-40, doi: 10.1109/29.1486.

[7] M. Basseville, "Detecting Changes in Signals and Systems – A Survey," Automatica, vol. 24, issue 3, May 1988, pp. 309-326, doi: 10.1016/0005-1098(88)90073-8.

[8] C. Truong, L. Oudre, and N. Vayatis, "Selective Review of Offline Change Point Detection Methods," Signal Processing, vol. 167, Feb 2020, pp. 1-20, doi: 10.1016/j.sigpro.2019.107299.

[9] H. Kim, K. Kim, S. Park, H. Kim, and H. Kim, "CoSimulating Communication Networks and Elecrical System for Performance Evaluation in Smart Grid," Applied Sciences, vol. 8, issue 1, Jan. 2018, pp. 85-109, doi:10.3390/app8010085.

[10] "NIST Framework and Roadmap for Smart Grid Interoperability Standards, Release 3.0," NIST special publication 1108r3, NIST, Sep. 2014, doi: 10.6028/NIST.SP.1108r3.

[11] "Guidelines for Smart Grid Cybersecurity, Volume 1 – Smart Grid Cybersecurity Strategy, Architecture, and High Level Requirements," NISTIR 7268 Revision 1, NIST, Sep. 2014, doi: 10.6028/NIST.IR.7628r1.

[12] H-J. Liao, C-H. R. Lin, Y-C. Lin, and K-Y. Tung, "Intrusion Detection System: A Comprehensive Review," Journal of Network and Computer Applications, vol. 36, issue 1, Jan 2013, pp. 16-24, doi: 10.1016/j.jnca.2012.09.004.

[13] H. Sedjelmaci and S. M. Senouci, "Smart Grid Security: A New Approach to Detect Intruders in a Smart Grid Neighborhood Area Network," Int. Conf. on Wireless Networks and Mobile Communications (WINCOM), 2016, pp. 6-11, doi: 10.1109/WINCOM.2016.7777182.

[14] W. Yu, D. Griffith, L. Ge, S. Bhattarai, and N. Golmie, "An integrated detection system against false data injection attacks in the smart grid," Security and Communication Networks, Wiley and Sons, vol. 8, issue 2, Jan. 2015, pp. 91-109, doi: abs/10.1002/sec.957.

[15] M. Ozay, I. Esnaola, F. T. Yarman Vural, S. R. Kulkarni, and H. V. Poor, "Machine Learning Methods for Attack Detection in the Smart Grid," IEEE Transactions on Neural Networks and Learning Systems, vol. 27, issue. 8, Aug. 2016, pp. 1773-1786, doi: 10.1109/TNNLS.2015.2404803.

[16] K. Smagulova, and A.P. James, "Overview of Long Short-Term Memory Neural Networks," In: James A. (eds) Deep Learning Classifiers with Memristive Networks. Modeling and Optimization in Science and Technologies, Springer, vol 14, Apr. 2020, pp139-153, doi:10.1007/978-3-030-14524-8_11.

[17] J. Toubeau, J. Bottieau, F. Vallée and Z. De Grève, "Deep Learning-Based Multivariate Probabilistic Forecasting for Short-Term Scheduling in Power Markets," in *IEEE Transactions on Power Systems*, vol. 34, no. 2, pp. 1203-1215, March 2019, doi: 10.1109/TPWRS.2018.2870041.

[18] B. Palmintier, D. Krishnamurthy, P. Top, S. Smith, J. Daily and J. Fuller, "Design of the HELICS high-performance transmission-distribution-communication-market co-simulation framework," Workshop on Modeling and Simulation of Cyber-Physical Energy Systems (MSCPES), 2017, pp. 1-6, doi: 10.1109/MSCPES.2017.8064542.

[19] G. F. Riley and T. R. Henderson, "The ns-3 Network Simulator," In: Wehrle K., Güneş M., Gross J. (eds) Modeling and Tools for Network Simulation. Springer, doi: 10.1007/978-3-642-12331-3_2.

[20] *GridLAB-D Simulation Software*. Available: https://www.gridlabd.org/[retrieved:May,2020].

[21] D. P. Chassin, K. Schneider, and C. Gerkensmeyer, "GridLAB-D: An open-source power systems modeling and simulation environment," 2008 IEEE/PES Transmission and Distribution Conference and Exposition, Chicago, IL, 2008, pp. 1-5, doi: 10.1109/TDC.2008.4517260.

[22] *MATLAB - MathWorks - MATLAB & Simulink*. Available: https://www.mathworks.com/products/matlab.html/ [retrieved: May, 2020].

The Importance of Accurate Battery Models for Power Assessment in Smart Energy Systems

Yukai Chen, Sara Vinco
Department of Control and Computer Engineering, Politecnico di Torino
Corso Duca degli Abruzzi, 24, 10129 Torino, Italy
Email: yukai.chen@polito.it, sara.vinco@polito.it

Abstract—**The smart energy system is characterized by a broader combination of various energy sources and energy storage devices with smart control management and increased attention to optimization for increasing energy efficiency. The fundamental dimension in the smart energy system design is the power assessment of the possible design architecture. This demand imposes a need for accurately tracking the system's power flow, simulating and validating the system's behavior, and applying additional optimization and exploration during the design time. Thus, it is evident that simulation is a critical step in the design flow of a smart energy system. One essential element to enable such accurate simulation is the precise model of the power generation and consumption. While sophisticated models for energy sources exist, the power flow in the system does not perfectly match the power drawn from the energy storage devices because the battery, as the primary energy storage device in the smart energy system, has non-ideal discharge characteristics. We propose adopting an elaborate battery model for the smart energy system's accurate power assessment in this work. We show the importance of battery model accuracy when conducting a power assessment using two different case studies.**

Keywords–Power Modeling and Simulation; Battery Modeling; Smart Energy System; Design Time Optimization; Energy Optimization.

I. Introduction

The smart energy system contains different scales of energy systems executed by the smart control management policy. It covers applications from watts to kilowatts levels, from small-scale Internet of Things (IoT) nodes to large-scale smart grid applications. High energy efficiency is a critical requirement for the small-scale smart energy systems since they only have a small-size energy storage device and power source component. Some of them even do not install the power source component. A typical small-scale example is a battery-powered IoT sensor node. Maximizing this kind of sensor node's lifetime is critical during the design time since the nodes are distributed in a very dispersed manner and the cost of replacement is very high. For large-scale smart energy systems, the adoption of green power sources and energy storage devices is considered a promising solution to reduce the impact on the environment and save users' costs. Therefore, these kinds of large-scale smart energy systems need a careful design to overcome the oscillating nature of the harvested environmental quantities and the users' undisciplined power consumption to achieve an optimal balance between power generation and consumption.

To improve the smart energy system's energy efficiency, the designer must assess the system's real-time power flow during the design time using computer-aided modeling and simulation tools. The traditional approach in the academic and industry fields to designing a smart energy system relies on a model-based methodology that depends on existing models provided by commercial simulation tools like Matlab/Simulink. Commercial tools have enhanced features and functionalities, and they are ordinarily user-friendly. Still, there are several limitations of the commercial tools: (1) it is hard to extend them as proprietary tools; (2) it is difficult to extend the performance of the built-in classic models to generate more accurate results; (3) commercial tools are not designed for efficiently and simultaneously simulating the physical portion (usually continuous-time) and the cyber portion (usually discrete-time) of the systems, while these two portions are the intrinsic features coexisting in the smart energy system; and (4) a smart energy system is a typical heterogeneous system composed of various components, with different parts possibly needing different models of computations in the simulation; commercial tools and platforms lack this kind of heterogeneous modeling and simulation support.

To tackle these limitations, several methodologies have been proposed in the literature that borrowed the existing approaches from the domain of electronic system design. One common feature shared by these solutions is building a database filled with pre-characterized models of the system's various components. This feature does allow the designer to select a model of an element from the pre-defined model database by assuming a given level of abstraction and a given semantics of the model. It speeds up modeling the whole smart energy system; however, one critical issue is ignored during the power assessment. The smart energy system is typically composed of three main categories: power source components, energy storage components, and power consumption components. When the battery plays the power source role in the system, the power consumed by the load does not have a perfect 1:1 match with the power provided from the battery, as it is not an ideal electric device. The reason for the mismatch is that the actual power delivered by the battery depends on its current State Of Charge (SOC), the current magnitude, and the load frequency. More specifically, the higher the current amplitude and load frequency, the more power is being consumed from the battery. Hence, under the same conditions, a larger discharge current or high-frequency discharge current profile will consume more battery capacity.

To address this issue, in this paper, we propose to incorporate one circuit equivalent battery model, which has SOC and dynamic current dependency characteristics, in the power simulation of a smart energy system. We adopt an elaborated battery model to simulate the smart energy system for tracking accurate real-time power flow in the system. Two case studies relative to different smart energy system scales are simulated in our work to illustrate the importance of an accurate battery model for power/energy assessment in various smart energy system applications.

The rest of the paper is organized as follows. Section II provides the required background and the motivation of this work; Section III describes the battery model adopted in our work; Section IV reports the simulation framework used in our experiments; Section V shows two case studies to indicate our proposal; finally, Section VI concludes the paper.

II. Background and Motivation

A large number of research activities are currently devoted to optimizing energy efficiency in the smart energy system. A high energy-efficient design relies on an accurate power/energy analysis, and modeling and simulation of the smart energy system are using widely approved techniques to conduct power/energy assessment during the early design time. Several approaches for modeling and simulation of smart energy systems have been proposed in the literature to address different application contexts, ranging from general-purpose electrical energy systems [1], small-scale IoT devices [2], large-scale system like smart grids [3], and medium-scale systems as Electric Vehicles (EV) [4]. Several different popular simulation approaches are introduced in the following paragraphs.

Hardware-in-the-loop approaches mix software simulated models with sensors and actuators or integrate power electronic devices such as inverters to test the combination of new technology in a controlled environment [5]. The resulting accuracy is higher than software simulation, but applications are restricted to small- and medium-scale smart energy systems. Proprietary tools, such as Simulink, are usually considered the de-facto standard. However, they are proprietary tools, thus not easily extensible and accessible. The designer can choose among several predefined components or instead implement his/her designs by relying on the provided libraries, which restricts the chances of developing custom component libraries and evaluating alternative models. Equation-based approaches, such as Modelica, decompose the system into elementary components, modeled with fundamental physics equations, or with predefined models [6]. This method restricts the kind of descriptions supported by each simulation infrastructure. It does not allow designers to model the cyber portion of a smart energy system effectively. Co-simulation approaches simulate specific aspects of the energy systems in their native environment, combined with other tools to estimate, e.g., the impact of network latency on control policies or the application of electricity rates [7]. This co-simulating mechanism leads to a very time-demanding and error-prone process for integrating components implemented with different characteristics, e.g., with discrete-time or continuous-time behaviors. Additionally, co-simulation moves the focus from smart energy system design to its interaction with other domains, thus not accurately reproducing the energy system components' behavior. It also lacks a big picture estimation of the power assessment in the smart energy system.

The main limitation of the previously presented approaches is that support for smart energy system modeling is limited, either in the kind of models or in the scale of supported systems. To overcome these limitations, the work in [2] targets more comprehensive support for the smart energy system in a SystemC-based framework, thus avoiding the integration of various tools and allowing the application of the methodology to a wide range of component models. Previous attempts in the literature have been made to adopt the standard SystemC simulation framework in the smart energy system. [8] uses SystemC for abstracting and modeling physical behaviors. However, the support for the power domain is limited to high-level waveforms or physical equations. It is also limited to the sole DC domain, and the modeling of the environment or physical evolution is restricted to input traces. The work in [9] takes inspiration from [2], but extends the support for cyber-physical electrical energy systems, targeting the Alternating Current (AC) domain and more accurate modeling of those physical aspects that profoundly affect power production and consumption.

Although previous works solve the unified simulation of AC and Direct Current (DC) domains of the smart energy system, there is one critical point that is ignored in the earlier works when conducting a simulation. The battery is usually the energy storage component in the current smart energy system, from small-scale to large-scale applications. Simultaneously, it is not an ideal electric device because the power drawn from the battery is not entirely equal to the power consumed by the load. The delivered power from the battery is strongly dependent on the SOC of the battery and the load current profile [10]. Especially in the smart energy system, the power load consumption is typically non-stationary, e.g., the electric motor's load consumption in the EV and all the household appliances in the home. The simple battery model directly provides the same power quantity as the load requests, but it is not the battery's natural discharge characteristics. To compensate for this defect in the smart energy system's power simulation, we need a battery model sensitive to its SOC and the current load profile. Otherwise, the power simulation results cannot generate the accurate power/energy flow in the smart energy system, which leads to an overly optimistic power assessment because the battery is treated as an ideal electric device. Therefore, in this work, we propose incorporating one circuit equivalent battery model with SOC and load current dependence features in the power simulation to obtain an accurate power assessment in the smart energy system.

III. Battery Modeling

The battery is the most popular energy storage device from small-scale to large-scale smart energy systems due to its excellent performance on power delivery capability and high energy storage density. Therefore, the elaborated battery model plays a vital role in the design and optimization of smart energy systems. For instance, battery-powered IoT sensor nodes lifetime estimation and EV driving range prediction are possible only through advanced battery models and accurate simulation techniques. According to the various battery modeling techniques in the literature, there are three main battery models categories: : mathematical models, electrochemical models, and electrical equivalent circuit models. Although the mathematical models and electrochemical models have good accuracy, the complexity of the parameter identification and heavy computation are not convenient for the power assessment in the design time. An electrical equivalent circuit battery model is an excellent choice for power simulation of a smart energy system due to its high accuracy, accessible parameters identification, and light computation. An electrical equivalent circuit battery model is composed of resistors, capacitors, and voltage or current sources. Fig. 1 shows a widely used circuit equivalent battery model, which is considered as a sort of standard in the electronic design field because of its relatively good trade-off between simplicity and accuracy.

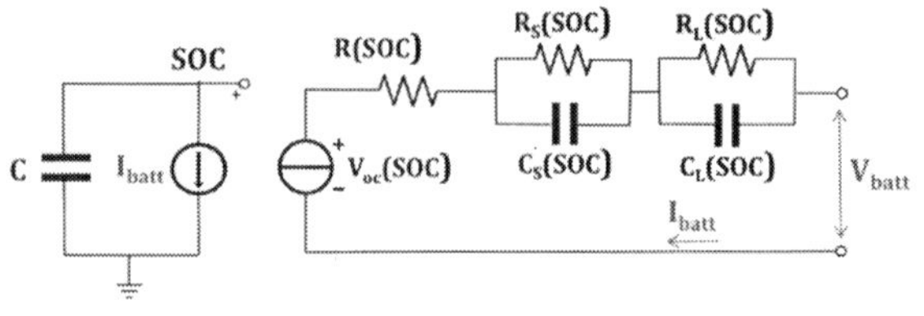

Figure 1. The widely used circuit equivalent battery model [11].

The left part of the model in Fig. 1 includes a capacitor C representing the nominal capacity and a current generator modeling the discharge current I_{batt}. The voltage across the capacitor tracks the SOC (node SOC). In the right part, a voltage-controlled generator expresses the dependence of battery open-circuit voltage V_{oc} on SOC. The Resistor-Capacitor (RC) network models the battery impedance, the series resistance $R(SOC)$ represents the internal resistance, and the two RC pairs track the short-term (R_S, C_S) and long-term (R_L, C_L) time constants of an instant response. Notice that all these parameters are, in the most general scenario, a function of the SOC. The model can track the battery voltage V_{batt} over time for load current profiles with different dynamics (in both time and frequency domains), even if they have the same average current values. In terms of tracking the SOC, the current generator I_{batt} on the left side will give the same result among different load current profiles with the same average current values as indicated in [12], which is not the expected behavior because of the *Rated Capacity effect* exists in a battery. *Rated Capacity effect* is a well-known non-ideal property of a battery. It reveals that the usable capacity depends on the magnitude of the discharge current: a battery is less efficient in converting its chemically stored energy into electrical energy if the current is large.

To integrate the *Rated Capacity effect* in the battery model shown in Fig. 1, the previous work [13] proposed to include this effect by adding a voltage generator $V_{lost}(I_{load})$ in series to the left part of the model (see Fig. 2). With this addition, SOC can be tuned to become more sensitive and accurate to different current magnitudes. However, the model cannot observe the difference between two current square-wave load profiles with the same average and swing values, but different frequencies. The frequency of the load profile is an important aspect; it can affect the SOC of battery like the current magnitude and is always ignored when conducting a power assessment of the energy system. Intuitively, a higher frequency load depletes the battery more since low-efficiency electrochemical reactions occur at a higher frequency. This dependency on load current frequency is underrated in previous works, while this effect is not negligible, as illustrated in [12]. This frequency dependency should be considered in the power analysis since the smart energy system's load profile is always irregular and unpredictable. The different frequency components are hidden in the current load profile, affecting the available battery charge.

For this purpose, another circuit equivalent battery model is proposed, as shown in Fig. 2. The frequency dependence is modeled by integrating the voltage generator $V_{lost}(f_{load})$ on

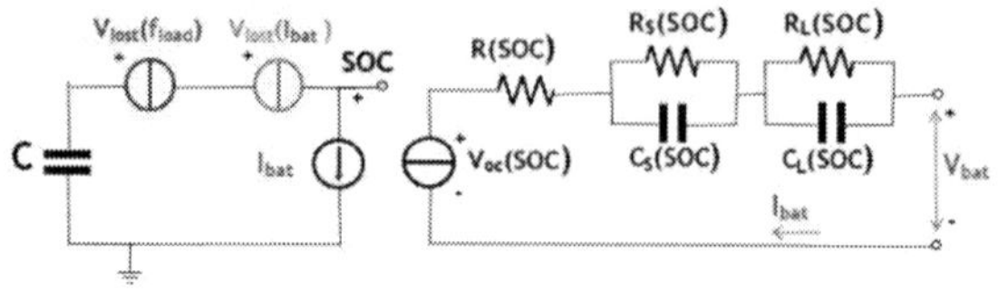

Figure 2. The circuit equivalent model sensitive to current dynamics [12].

the left side of the circuit, which will cause a voltage drop (a loss of SOC) depending on the frequency of the current load profile. In this work, we incorporate the battery model in Fig. 2 for the power assessment of the smart energy system. For a given battery, the methodologies to derive the dependencies of battery capacity on current load magnitude and frequency are used on the left side of the battery model in Fig. 2 and are described in [12]. It also introduces how to use the model in the simulation, particularly for the battery frequency-dependent simulation. The battery model shown in Fig. 2 represents one single battery cell; it can be directly adopted in the small-scale smart energy system with only one battery cell. However, for the battery energy storage system installed in the large-scale smart energy systems, the systems typically have a battery pack composed of massive cells connected in series and parallel to achieve expected power and energy rates. A commonly used approach to build the battery pack model assumes all the cells behave identically within the pack, then ideally scale up the battery cell's electrical parameters according to the serial and parallel connectivity. This method guarantees a faster simulation and higher flexibility in modeling and simulation of a large battery pack. In this way, not all the cells have to be simulated individually; a battery pack behaves at the electrical terminals the same way as a single battery cell, while the electric values have been ideally scaled-up.

IV. SystemC-AMS Simulation Framework

A. Simulation Framework Architecture

To conduct the smart energy system's power assessment, the designer needs to run a smart energy system simulation to track power production, power distribution, and power consumption. We can classify four main categories in the smart energy system: power source, power load, energy storage, and conversion components. The small-scale systems typically operate in the DC domain, while both DC and AC domains are covered for the large-scale system. Fig. 3 shows a generic architecture template of the smart energy system; it comprises all kinds of components included in the smart energy systems from DC to AC domain.

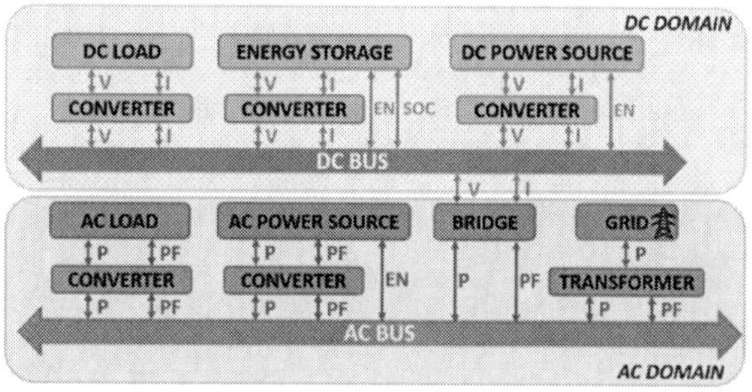

Figure 3. Generic smart energy system modular architectural template.

We adopted the simulation framework proposed in [2] in this work, although it only focuses on small-size smart electronic systems and targets various non-functional properties. In our work, we concentrate on the power property and extend the support of small-size smart electronic systems to the general smart energy system, as shown in Fig. 3. The adopted simulation framework indicated in Fig. 3 shows a bus-based modular architecture; it mimics the structure of a typical smart energy system. Notice that the small-scale system only works in the DC domain. Four main components are included in this framework, loads (acting as energy consumers), power sources (acting as power generators), converters, and energy storage devices (storing energy from power generators and providing to energy consumers).

The ports of the modules in the DC and AC domains shown in Fig. 3 are different, where the DC ports are modeled as V and I and the AC ports are modeled as P and PF. The reason is that the AC domain's power is not the same as the power in the DC domain. The power can be directly modeled as voltage and current in the DC domain. However, the AC power is composed of two components, namely, active and reactive power. The sum of these two powers is called apparent power. In general, the current and voltage in the AC domain are not the same sinusoidal curves; the two curves may be out of phase by a degree depending on the type of component. The PF is called the power factor, which is the parameter that represents this degree. Therefore, we can use P to describe the active power and the corresponding power factor PF to compute the apparent power.

Components are connected through a power bus in both AC and DC domains, which allows the power to combine and propagate within the system. The power management policy is implemented in the system buses (either AC or DC). Additionally, the policy considers the power balance between generation and consumption, to activate or de-activate components through enabling ports, as indicated by EN in Fig. 3. The connections of different components to the buses need converters, to maintain the same voltage level of the bus, or inverters, to convert between the DC and the AC domains. Connections between different domains are finally managed by bridges that behave the same as an inverter.

B. Implementation Simulation Framework

As described in the previous sections, a smart energy system is typically composed of various components. Heterogeneity is one main feature of the smart energy system. Manipulating different heterogeneous models in one system is not an easy task. We adopted the methodology proposed in [9] to model different components in the smart energy system by using SystemC-AMS language, which is also the reference language to implement the whole simulation framework. It supports multiple abstraction levels for the modeling and simulation in a wide range of domains by using a single language, which satisfies the heterogeneous modeling and simulation of the smart energy system. Models can be built by choosing the most suitable abstraction level, and native converters can be exploited to simulate different abstraction levels simultaneously.

SystemC is an extension of C/C++ language with specific libraries to describe hardware, and it is widely used in digital design for early-stage analyses. Its Analog Mixed Signal (AMS) extension [14] was invented for modeling and simulating the interactions between analog/mixed signal systems.. It provides three different Models of Computation (MoC) to cover various domains. Timed Data-Flow (TDF) models are scheduled statically by considering their producer-consumer dependencies in the discrete-time domain. Each TDF module is characterized by a simulation time step used by the TDF solver to insert timed activation events in the standard SystemC event queue. This event queue ensures efficient computation, as it avoids any runtime dynamic event management. Another two abstraction levels support continuous-time models. Linear Signal Flow (LSF) supports continuous-time modeling through a library of pre-defined non-conservative primitive modules that can be used for modeling the sophisticated mechanical model, e.g., the wind turbine model. Electrical Linear Network (ELN) MoC model the electrical network by connecting the instantiations of pre-defined primitives, which is used for modeling the circuit equivalent model, e.g., our adopted circuit equivalent battery model.

V. Simulation Results of Two Case Studies

We use two case studies to illustrate the importance of the elaborated battery model for the power assessment in the smart energy system. All simulations conducted in this section are implemented with SystemC-AMS version 2.3 and run on a server with Intel Xeon 2.40 GHz CPU and 128GB RAM, as well as Ubuntu operating system version 18.04.

A. Small-scale smart energy system case study

One multi-sensor IoT device described in [15] is selected as the first small-scale case study. It has (1) four sensors to monitor four different environmental quantities, namely, wind speed and direction, gas, PM2.5 and infrared; (2) one microcontroller that manages the scheduling of the sensors activities by controlling their power state; (3) one transmission unit for sending data to the data center, and (4) one lithium battery that provides the power to the device. This case study only operates in the DC domain, and there is only power consumed and energy storage components in the system. Fig. 4 shows the skeleton of this small-scale smart energy system.

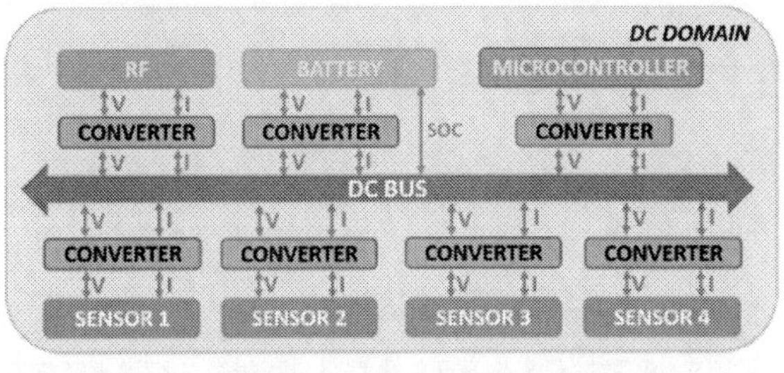

Figure 4. Modular architecture of smart IoT sensor nodes energy system.

All the components in the system are modeled by SystemC-AMS. The sensors, Radio Frequency (RF) transceiver, and microcontroller components are modeled by TDF MoC since they are pure power consumed units. Table I lists the power consumption of these units in both active and idle states. The battery is modeled by ELN MoC, as shown in Fig. 2. We selected the Panasonic 18650B lithium battery in our simulations. It has 3,200 mAh nominal capacity and 3.6V rated voltage. All the converters in the system are modeled by the same methodology proposed in [4].

This smart system's operating scenario is composed of a periodic sequence of the following tasks: sensing, computation,

TABLE I. POWER CONSUMPTION OF EACH COMPONENT IN THE SYSTEM.

Component	State	Voltage (V)	Current (mA)
Infrared Sensor	Idle	2.5	0.0001
	Active	5.0	10.0
Wind Sensor	Idle	5.0	0.001
	Active	12.0	50.0
Gas Sensor	Idle	2.0	0.0015
	Active	5.0	168.9
PM2.5 Sensor	Idle	2.0	2.0
	Active	5.0	220.0
RF	Idle	0	0
	Active	1.8	18.8
Microcontroller	Idle	3.0	0.002
	Active	3.0	6.0

and transmission. When the system executes these tasks, it is in the active period (T_{active}), then the system enters a longer idle period (T_{idle}) after these operations. The executive order of different components within the T_{active} interval becomes a critical point for improving the system's energy efficiency. Notably, different scheduling of sensors generate different load current profiles, finally affecting the available SOC of battery, which determines the lifetime of the system.

To show the influence of the battery model accuracy, we tested two scheduling policies of these four sensors. One is all the sensors work concurrently; the other is the sensors operate individually one by one. Both schedules are simulating with the simple battery model shown in Fig. 1 and the elaborated battery model indicated in Fig. 2. We set the sensing tasks to spend 10s individually, the computational task and transmitting task both execute for 5s, and the total length the working period is 120s. We run the simulations in a one-day length to compare the difference between two battery models and two schedules.

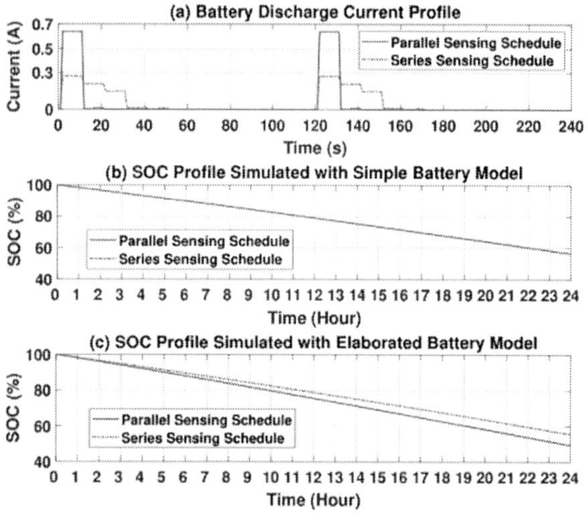

Figure 5. Results from two battery models with different schedules.

Fig. 5-a shows the battery discharge current profile within two periods for both schedules; it indicates that the parallel schedule has higher peak currents; thus, it should generate more significant *Rated Capacity effect* on the battery. However, the battery model shown in Fig. 1 cannot show this effect as illustrated in Fig. 5-b, the SOC profiles within one day of these two schedules are the same. We can see the difference between these two schedules clearly by using the more elaborate battery model in the simulations, as shown in Fig. 5-c. This reveals that the parallel schedule consumes 6.14% more than the series sensing schedule. Designing the optimal energy-efficient schedule is not the target of this work; our aim is to illustrate that a more comprehensive battery model allows exploring the optimal schedule during the design time accurately.

B. Large-scale smart energy system case study

The second case study is a large-scale smart energy system composed of is an EV, a Photovoltaic (PV) array, a house, and the utility grid. The architecture of this case study is shown in Fig. 6 and it covers both the DC and AC domains. The PV array is modeled by starting from a single PV module model [16], then ideally scaling up to the size of the PV array with 15 300W rated power PV modules. The single battery cell model is the same as in our previous case study. It scaled up to a 30kWh EV battery pack model according to a 50s50p configuration. The EV motor model is derived by the method provided in [4]. The power consumption of the house is extracted from the dataset [17]. The input traces of solar irradiance used in the simulation are extracted from the dataset provided by the National Renewable Energy Laboratory (NREL) Measurement and Instrumentation Data Center (MIDC) [18]. For the driving profile, we assumed the EV operates a daily commute routine. The grid is modeled by a module that can absorb the system's surplus power and provide the system's power deficit.

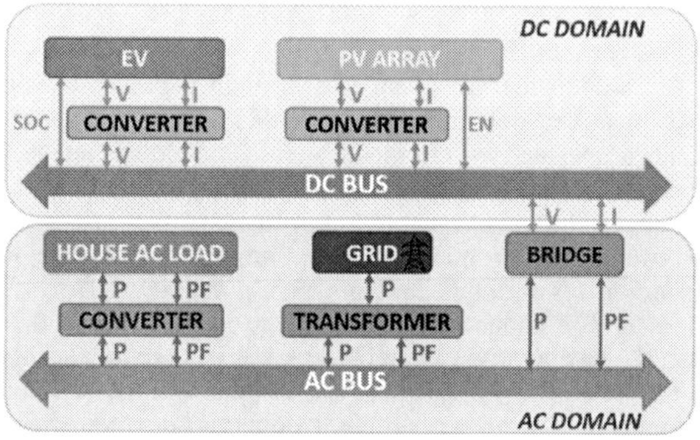

Figure 6. Modular architecture of smart house energy system.

The EV operates the daily commute routine between the house and the working place. The operational scenarios of this case study have two phases:1) when the EV is not plugged in at the house, the PV array provides the solar power to support power consumption of the house; if any additional power, it will be sold to the grid; any power deficit will be bought from the grid; 2) when the EV is connected to the house, the power consumption of the house is provided by the EV first, then the house starts to buy power from the utility grid if the battery pack is depleted; finally, the EV and the house start to buy the power from the grid during the lowest electricity price period. A time-dependent electricity price is adopted in the simulation, as indicated in Table II. These are the basic operations of this case study. Smart management policies can be explored in this case study in order to improve the efficiency of energy or for cost saving. While this is beyond the target of this work, we aim to show the accuracy improvement of the power assessment after incorporating an elaborated battery model.

Fig. 7 shows the EV battery pack SOC evolution for the weekdays in one week. Fig. 7-a and Fig. 7-b show the

TABLE II. ELECTRICITY PRICES FOR DIFFERENT TIMES OF THE DAY.

Price Category	Value ($/kWh)	Time span
Buying F1	0.220	10am-3pm 6pm-9pm
Buying F2	0.215	7am-10am 3pm-6pm 9pm-11pm
Buying F3	0.200	11pm-7am
Selling	0.030	all day

power consumption and generation in the system. Fig. 7-c compares the SOC profiles derived from two different battery models. This reveals that the battery pack has full charge at the beginning; then, it starts to discharge power after the EV leaves the house to the working place; the SOC remains stable when the EV is parked at the working place; then, the SOC decreases again when the EV returns the house; the EV provides the power to the house if needed; finally, the battery pack is charged after 11 pm since the electricity price decreases to the lowest price. We assume the constant speed of the EV during the driving period, and we focus on the period when the EV interacts with the system.

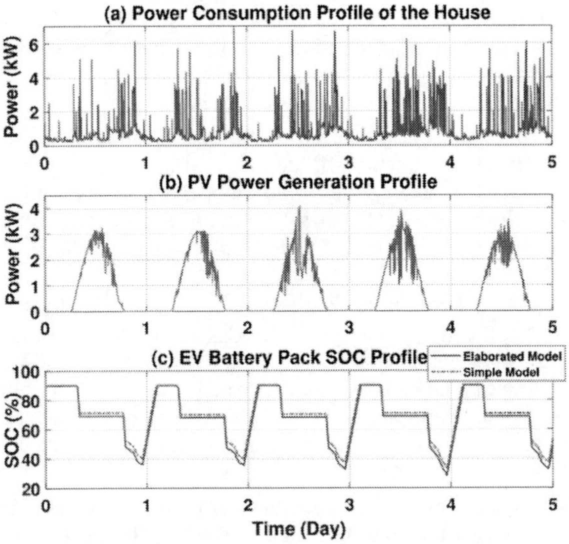

Figure 7. Simulation results from two different battery models.

This case study verifies the importance of the battery model accuracy in the power simulation of a smart energy system again. Fig. 7-c illustrates that the simple battery model makes an underestimated power assessment during the simulation; the SOC profile derived by the elaborated battery model is below the simple battery model one, which means the battery pack needs to be charged more after 11 pm. The simple battery model leads to an optimistic inaccurate power estimation of the smart energy system. For a long term comparison, the total bought energy from the grid after one year is 4,163 kWh for simulating with the elaborated model and 3,882 kWh with the simple model, which has a 7.2% difference.

VI. CONCLUSION

An accurate power assessment in the smart energy system simulation requires sophisticated models of different components in the system, and the battery model is the critical one due to its non-ideal properties. This paper proposed incorporating one elaborated circuit equivalent battery model that accounts for SOC and current load dynamics in the power simulation of the smart energy system. The elaborated model can conduct a more accurate power assessment of the smart energy system than using the traditional circuit equivalent battery model. We demonstrate the importance of the battery model accuracy for the power estimation of the smart energy system by using two different scales of case studies.

REFERENCES

[1] S. Yue, D. Zhu, Y. Wang, M. Pedram, Y. Kim, and N. Chang, "SIMES: A simulator for hybrid electrical energy storage systems," in International Symposium on Low Power Electronics and Design (ISLPED). IEEE, 2013, pp. 33–38.

[2] S. Vinco, Y. Chen, F. Fummi, E. Macii, and M. Poncino, "A layered methodology for the simulation of extra-functional properties in smart systems," IEEE Transactions on Computer-Aided Design of Integrated Circuits and Systems, vol. 36, no. 10, 2017, pp. 1702–1715.

[3] M. A. Mohamed and A. M. Eltamaly, Modeling and simulation of smart grid integrated with hybrid renewable energy systems. Springer, 2018.

[4] Y. Chen et al., "A systemc-ams framework for the design and simulation of energy management in electric vehicles," IEEE Access, vol. 7, 2019, pp. 25 779–25 791.

[5] B. Palmintier et al., "A power hardware-in-the-loop platform with remote distribution circuit cosimulation," IEEE Transactions on Industrial Electronics, vol. 62, no. 4, 2014, pp. 2236–2245.

[6] P. Palensky, E. Widl, and A. Elsheikh, "Simulating cyber-physical energy systems: Challenges, tools and methods," IEEE Transactions on Systems, Man, and Cybernetics: Systems, vol. 44, no. 3, 2013.

[7] P. Palensky, A. van der Meer, C. Lopez, A. Joseph, and K. Pan, "Applied cosimulation of intelligent power systems: Implementing hybrid simulators for complex power systems," IEEE Industrial Electronics Magazine, vol. 11, no. 2, 2017, pp. 6–21.

[8] J. M. Molina, X. Pan, C. Grimm, and M. Damm, "A framework for model-based design of embedded systems for energy management," in 2013 Workshop on Modeling and Simulation of Cyber-Physical Energy Systems (MSCPES). IEEE, 2013, pp. 1–6.

[9] Y. Chen, S. Vinco, D. J. Pagliari, P. Montuschi, E. Macii, and M. Poncino, "Modeling and simulation of cyber-physical electrical energy systems with SystemC-AMS," IEEE Transactions on Sustainable Computing, 2020.

[10] D. Baek et al., "Battery-aware operation range estimation for terrestrial and aerial electric vehicles," IEEE Transactions on Vehicular Technology, vol. 68, no. 6, 2019, pp. 5471–5482.

[11] M. Chen and G. Rincón-Mora, "Accurate electrical battery model capable of predicting runtime and IV performance," IEEE Transactions on Energy Conversion, vol. 21, no. 2, 2006, pp. 504–511.

[12] Y. Chen, E. Macii, and M. Poncino, "A circuit-equivalent battery model accounting for the dependency on load frequency," in Design, Automation & Test in Europe Conference & Exhibition (DATE), 2017. European Design and Automation Association, 2017, pp. 1177–1182.

[13] L. Benini, G. Castelli, A. Macii, E. Macii, M. Poncino, and R. Scarsi, "Discrete-time battery models for system-level low-power design," IEEE Transactions on VLSI Systems, vol. 9, no. 5, 2001, pp. 630–640.

[14] IEEE, "Standard for Standard SystemC Analog/Mixed-Signal Extensions Language Reference Manual," Std 1666.1-2016, 2016, pp. 1–236.

[15] Y. Chen, D. Jahier Pagliari, E. Macii, and M. Poncino, "Battery-aware design exploration of scheduling policies for multi-sensor devices," in Proceedings of the 2018 on Great Lakes Symposium on VLSI, 2018, pp. 201–206.

[16] S. Vinco, Y. Chen, E. Macii, and M. Poncino, "A unified model of power sources for the simulation of electrical energy systems," in Proceedings of the 26th edition on Great Lakes Symposium on VLSI, 2016, pp. 281–286.

[17] J. Kelly and W. Knottenbelt, "The uk-dale dataset, domestic appliance-level electricity demand and whole-house demand from five uk homes," Scientific data, vol. 2, no. 1, 2015, pp. 1–14.

[18] National Renewable Energy Laboratory (NREL), "Measurement and Instrumentation Data Center," URL: https://midcdmz.nrel.gov/ [accessed: 2020-07-27].

Powertrain Modeling and Range Estimation of The Electric Bus

Donkyu Baek

School of Electronics Engineering
Chungbuk National University
Cheongju, Republic of Korea
Email: donkyu@cbnu.ac.kr

Abstract—**The driving range of Battery Electric Vehicles (BEVs) has been fairly extended during recent years, as a consequence of improvements in energy density of lithium-based batteries. As the scope of application of electric vehicles has expanded, electric motors have been used in trucks and buses, as well as simple passenger vehicles. There are several issues in electric buses for the decision of daily route and finding the optimal battery size. In this work, we propose and incorporate an electric bus powertrain model into a range estimator that takes into account slope, speed limits as well as traffic information. We introduce two case studies as applications of the proposed range estimator: (i) the fast route decision and (ii) the iteration-based bus battery sizing.**

Keywords–Electric vehicle; powetrain model; range estimator.

I. Introduction

Although most of the recent Battery Electric Vehicle (BEV) models have significantly extended the driving range (even less costly EVs can reach the 200-250 miles autonomy), range anxiety is still perceived as a major issue. This is due to both the limited battery performance, which largely depends on working and operating conditions, and the still lacking installations of charging stations, especially in Europe [1].

All BEVs include some form of a real-time driving range estimator based on battery State-of-Charge (SoC). Many researchers have addressed the issue of improving the estimator by accounting for all the possible factors contributing to energy consumption: among others, road topology and grade, speed, acceleration/deceleration, weather conditions, vehicle current location, use of on-board electric devices (e.g., A/C), and driving styles (e.g., normal vs. aggressive) [2][3].

As electric motors are used in various types of gasoline vehicles, electric buses have also been introduced on the market [4]. The bus is a good candidate for converting the gasoline powertrain system into an electric power system because of the motor's characteristics, such as being emission free, having low noise, having small vibration, being easy to maintain, etc. However, there are several issues with using electric buses in the public transportation system. We should know which route is more efficiency for the electric bus considering the road slope and the traffic. Also, for a given bus route, optimal battery sizing is another issue. Performing driving tests with different battery sizes is the most obvious way to find the optimal battery size, however, this is very time consuming.

In this work, we propose an electric bus powertrain modeling and simulation framework. We first implement several parameters and component models for a complex vehicle simulation system. Then, we implement an equation-form powertrain model to reduce runtime for energy simulation by extracting coefficients from complex simulation results. The powertrain model is incorporated in a range estimator that updates EV power consumption along with a given driving cycle characteristics including road slope, speed limit and traffic conditions.

The proposed fast electric bus powertrain model and range estimator are useful for runtime decision making and off-line battery sizing. We introduce two case studies as applications of the proposed range estimator: (i) the fast route decision and (ii) the iteration-based bus battery sizing.

The paper is organized as follows: Section II reports the related work; Section III describes the system models (i.e., powertrain, battery and route models) and the powertrain modeling process. Section IV reports simulation results and case studies. Finally, Section V draws some conclusions.

II. Background and Related Work

As a consequence of the worldwide increase in the number of BEVs, the automotive industry is facing some new challenges related to battery pack volume, weight, lifetime and cost. Furthermore, nowadays charging stations are not widespread in all geographical areas, and charging time is still too long with respect to the traditional refueling [5].

While driving range estimation is not a significant issue for Internal Combustion Engine Vehicles (ICEVs), it is more challenging for BEVs because some parameters largely affect the lithium-based battery pack energy at each charge/discharge cycle: current rate, temperature, and even driving style [6][7]. Despite the progress made in producing battery cells with similar energy yield at different discharge currents, depleting a battery at different rates generally leads to different total capacity (Ah) [8]. On top of that, the maximum battery energy decreases over time, even in case of non-connection to a load, as a consequence of deteriorating chemical processes [9].

There are many published papers addressing the issue of energy analysis and optimization in EVs. Most of these works leverage upon linear battery models [10]–[12] and, therefore, they do not include some important non-linear characteristics, such as the real dependency of battery voltage, current and efficiency to SoC. This non-linearity is sometimes just approximated [13] or described by a rather simplified mathematical model as in [14], where the authors proposed a steady-state (i.e., resistive) equivalent circuit. Three energy prediction methods are presented in [15]; however, the related framework should be simplified in order to have a more practical application. Recently, some papers suggest non-linear battery simulation working with EV driving simulation [16][17].

III. System Model and Estimation

A. Powertrain Model

The power consumption of an EV depends on body shape including facial area, curb weight, road slope and types of tires as well as on the speed and acceleration of the EV. Figure 1 shows the dynamic power by a motor rotating with torque T and angular velocity ω. Four resistances are acting on a vehicle, where F_R, F_G, F_I, and F_A are the rolling, gradient, inertia and aerodynamic resistances, respectively.

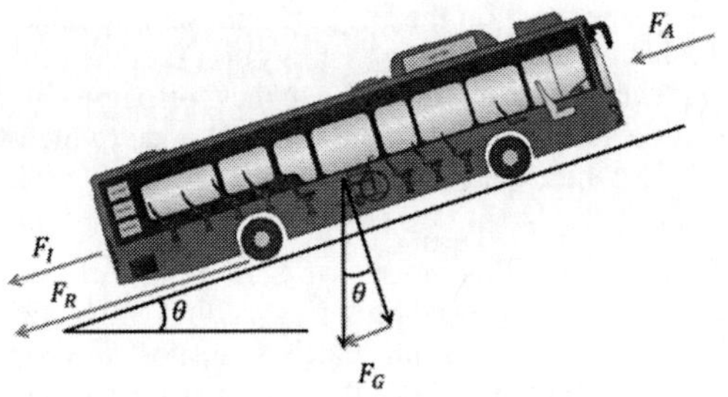

Figure 1. Forces on an EV.

The power consumption at the EV powertrain is the following [16][17]:

$$\begin{aligned} P_{dyna} &= T\omega = F\frac{ds}{dt} = Fv = (F_R + F_G + F_I + F_A)v \\ F_R \propto C_{rr}W,\ F_G &\propto Wsin\theta,\ F_I \propto ma,\ F_A \propto \frac{1}{2}\rho C_d A v^2 \\ P_{dyna} &\approx (\alpha + \beta sin\theta + \gamma a + \delta v^2)mv \end{aligned} \tag{1}$$

where C_{rr}, W, θ, m, v, a, C_d, A are the rolling coefficient, weight, road slope, vehicle mass, vehicle speed, acceleration, drag coefficient, and vehicle facial area, respectively. This relation is simplified as a function of four coefficients α, β, γ and δ.

The powertrain efficiency of the electric motor and drivetrain is less than 100%. The efficiency depends on the operating RPM (revolutions per minute) and torque when the EV drives. On top of that, the drivetrain mechanical movement causes a power loss while delivering power to the wheels. The following EV specific power model considers power losses by the motor and drivetrain [17]:

$$P_{EV} = P_{dyna} + C_0 + C_1 v + C_2 v^2 + C_3 T^2 \tag{2}$$

where C_0, C_1, C_2, and C_3 are the coefficients for constant loss, iron and friction losses, drivetrain loss, and copper loss, respectively.

Unlike ICEVs, the electric motor works like a power generator when EV reduces its speed. This is done by a regenerative braking system, which converts the kinetic energy on the wheel into electric energy and sends it to the battery. The power generation by regenerative braking is modeled by a function of the negative motor torque and vehicle speed, as follows:

$$P_{regen} = \epsilon T v + \zeta \tag{3}$$

where ϵ and ζ are regenerative braking coefficients.

B. Powertrain Modeling Process

There are several powertrain simulators in academia and industry. ADvanced VehIcle SimulatOR (ADVISOR) is one of well-organized vehicle simulators that takes into account various factors of vehicles including engines, electric traction motors, types of drivetrains, shape of chassis, etc. [18]. It is possible to implement a certain type of vehicle in ADVISOR by setting powertain parameters and simulating various vehicle driving environments by changing its powertrain or driving profile. Power consumption, battery state of charge and emission over time are simulated for a given driving cycle and vehicle setup.

ADVISOR, however, is not suitable to simulate power consumption in a live manner because of its relatively long runtime. ADVISOR considers overall vehicle dynamics and energy flow from torque on the wheels to the engine or battery pack. Overall vehicle simulation results show energy flow in detail, and this is important for energy analysis. However, it is not efficient to estimate the current load from the battery point of view and make a decision to find the optimal route or vehicle velocity.

Based on the arguments above, instead of using ADVISOR itself, we adopt the vehicle powertrain models from (1) to (3) and use ADVISOR for the extraction of model coefficients. Figure 2 shows the overall process for the electric bus characterization. The process consists of three phases: i) parameter extraction phase, ii) modeling phase and iii) simulation phase.

1) Parameter extraction phase: First, we choose a vehicle for the ADVISOR simulation. ADVISOR requires several parameters and models for the simulation (e.g. motor model, vehicle chassis model and battery model). Vehicle manufacturers officially unveil their vehicle specifications on their website, such as the maximum motor power and torque and the time to reach 100 km/h. This information is used to implement detailed parameters and models for ADVISOR. In this parameter extraction phase, we implement an electric motor model and a drivetrain efficiency model, and a battery model for the ADVISOR simulation. We implement a motor torque map from the maximum motor torque/RPM, the time to reach 100 km/h and vehicle curb weight. We then implement a motor efficiency map using the torque map, battery size and driving range. The drivetrain efficiency model is obtained from the driving range and resistances acting on a vehicle where we calculate resistances using the vehicle body shape and the type of tires. The battery model is easily obtained from battery architecture and the battery cell specification. These models are imported into the ADVISOR system and used to simulate the complex energy flow.

2) Modeling phase: ADVISOR simulates energy flow with an electric vehicle model obtained from Section III-B1 and a driving cycle. We perform simulations to obtain plentiful driving data with various vehicle speeds and road slopes. The simulation results include power consumption by vehicle speed and road slope over time. The driving cycles include driving on flat road with various vehicle speeds and accelerations on various road slopes. Test driving on various road slopes is also performed. We use a multi-variable linear regression method to extract coefficients of the powertrain models from (1) to (3) [17].

3) Simulation phase: The equation form powertrain model from (1) to (3) is used to extract the power consumption

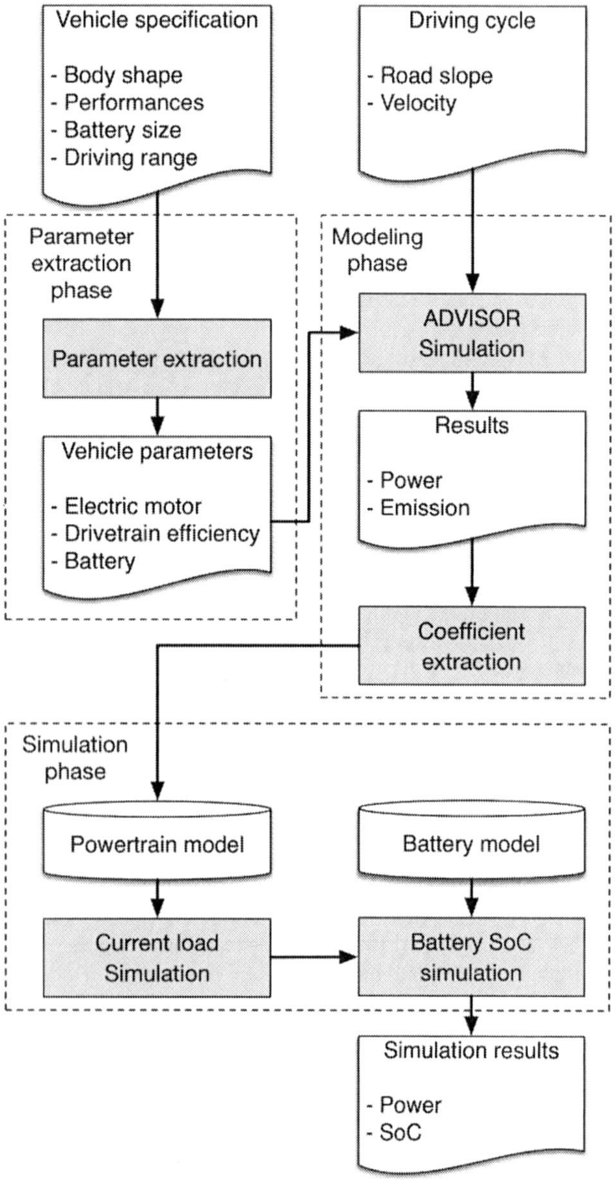

Figure 2. Overall process for the electric bus characterization.

by a given driving cycle promptly. We use the obtained power consumption to simulate the battery charging/discharging operation. The following Section III-C describes the details of battery SoC estimation.

C. Battery Pack Model

The EV battery pack typically includes a large number of Lithium battery cells. For example, a Tesla Model 3 battery pack consists of 444 Panasonic NCR18650B cells of 3,400 mAh nominal capacity with 74p6s arrangement. Hence, to model a pack, we need to model each individual cell, taking the load current and SoC variations of the usable battery capacity into account. This can be done with a circuit-equivalent model accounting for the capacity dependency on current magnitude and dynamics [19][20], as represented in Figure 3. The left-hand part of the circuit models the battery lifetime, with a capacitor C representing the battery storage capacity (Ah)

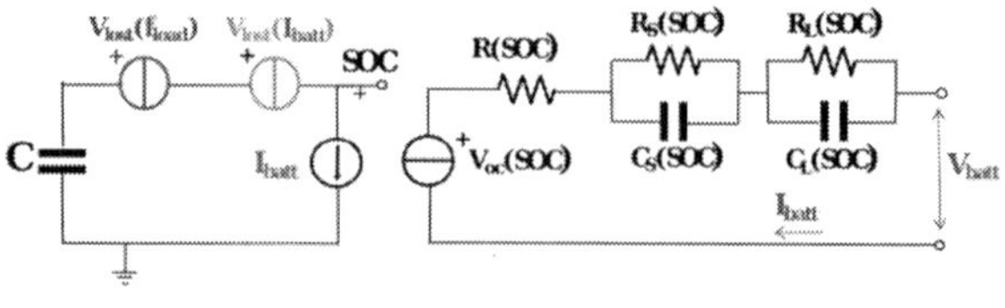

Figure 3. Circuit-equivalent model for battery cell [19].

and a current generator I_{batt} representing the battery current requested by the load. As the available capacity of the battery is affected by the load current values distribution, there are two voltage generators on the left part representing the dependency of the battery capacity (i.e., energy) on current values and the dependency on load current frequency, respectively. Both generators decrease the voltage at the SoC node (which represents the SoC) when either the current magnitudes or frequencies increase.

Starting from this model, we built a pack model by simply scaling parameters based on series/parallel connection. Besides its simplicity (e.g., cell mismatches are not considered) this is still a more realistic model than a linear one neglecting state-dependent information. Consideration of battery temperature is also a very important issue of battery SoC estimation. However, we leave the topic as a future work and focus on the state-dependent SoC estimation in this paper.

D. Driving Cycle Model

We extract a driving cycle, which represents a vehicle driving in a city. The driving cycle includes vehicle speed and road slope over time. We first extract a route to a destination and use related traffic information or rules and altitude information. We use speed limit and road traffic to synthesize the vehicle speed, whereas the altitude profile along the route is converted into road slopes. Figure 4 shows the road traffic and altitude of an example route from Google Maps [21] on the upper and lower subplots, respectively. Each color on the route means different levels of traffic: red means heavy traffic, orange means medium traffic, and blue means no traffic, respectively. We easily obtain the road slope from the change of altitude per distance unit.

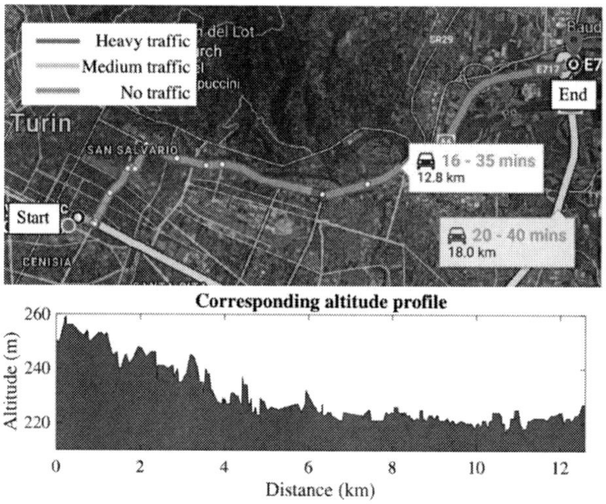

Figure 4. An example route on Google Maps [21] and its related altitude data.

We do not consider the acceleration/deceleration occurring in correspondence to speed changes. Such approximation is not critical however, because the acceleration energy is small compared with the total energy consumption of each segment: acceleration/deceleration seldom last more than a few seconds.

IV. Simulation Results

We implemented a powertrain model of a BYD K9 bus from the vehicle specification and experiment results [4][22][23]. The gross vehicle weight of BYD K9 is 18,000 kg and the facial area of K9 is 2.55 by 3.36 m. K9 includes two electric motors whose maximum motor torque and power are 350 Nm and 90 kW, respectively. The motor type is in-wheel BYD-TYC90A Brushless Permanent Magnet Synchronous Motor. The maximum RPM is 7,500 and the gear box ratio is 17.7:1. The manufacturers unveiled the driving range of K9 as being an average 250 km, based on their experience. Battery size is 320 kWh and the maximum road slope to climb is 15%.

A. *Vehicle Parameter Extraction*

We used [24] to set the parameters of the electric motor that Larson transportation institute tested on the BYD electric bus more than eight months (from August 29, 2013 to May 13, 2014). The institute reported the maximum acceleration of the bus from stop: 4.8 s to 10 mph, 9.0 s to 20 mph, 16.2 s to 30 mph, 32.4 s to 40 mph and 47 s top speed (43 mph), respectively. We extracted (a) the maximum motor torque map and (b) the maximum motor power by RPM, with repeataed ADVISOR simulations, as shown in Figure 5.

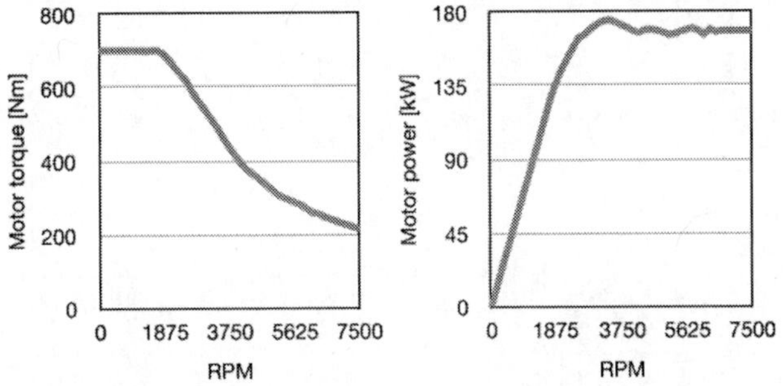

(a) Maximum motor torque map. (b) Maximum motor power.

Figure 5. Motor parameter extraction results. (a) is the maximum motor torque map and (b) is maximum power map.

We specified the detailed vehicle parameters as indicated in Figure 6; this figure also shows the ADVISOR user interface for parameter extraction. The motor, together with the efficiency and battery models, are imported into ADVISOR, and based on the simulation results, we set the parameters so that the simulation results follow the experimental results. Figure 7 shows the difference between experimental results and simulation results of the driving range. We set the parameters for drivetrain efficiency to follow the trend of driving range by vehicle speed. There are about 200 km range difference between two lines respecting the experimental driving range and the simulation driving range in Figure 7. However, the range trend by the speed is similar enough. Also, the range difference is resolved by updating the battery model, as explained in the following section.

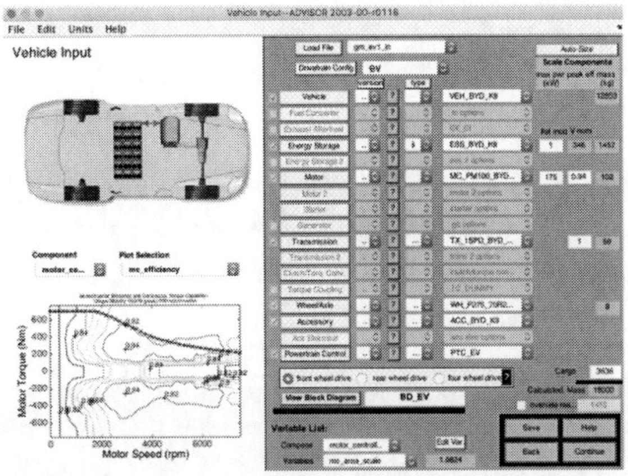

Figure 6. ADVISOR simulation setup.

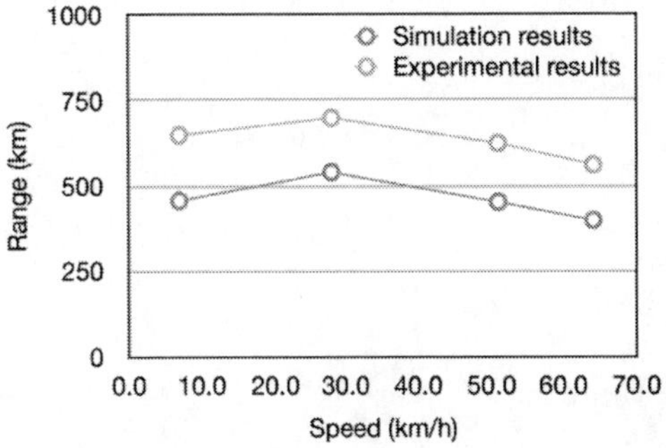

Figure 7. Driving range validation under efficiency parameter extraction.

B. *Vehicle Powertrain Modeling*

We extracted the coefficients of (1), (2) and (3) with a number of ADVISOR simulations. Table I summarizes the model coefficients of BYD K9. Figure 8 shows the difference between the estimation of power consumption by the vehicle simulator and the powertrain models; the normalized root-mean-square error is 9.12%.

TABLE I. MODEL COEFFICIENTS FOR BYD K9.

α	0.098	β	9.7562	γ	1.2016	δ	0.0001
C_0	1000.0	C_1	1378.2	C_2	0.00001	C_3	0.000015
ϵ	0.4095	ζ	2178.5				

C. *Vehicle Simulation Setup*

In our experiment, we followed the battery pack configuration provided by BYD to build our battery model. The battery is LiFePO4 (Lithium Iron Phosphate) with 540 V battery pack voltage. The battery pack consists of three battery modules and has 108 kWh capacity. We assumed that each battery cell in the pack is ideally balanced in the following experiments, then we built the battery pack model as indicated in Section III-A. Concerning the regenerative braking phase, we assume that charging efficiency is 20%, which means that 20% of the kinetic energy is converted to electric energy and transferred into the pack.

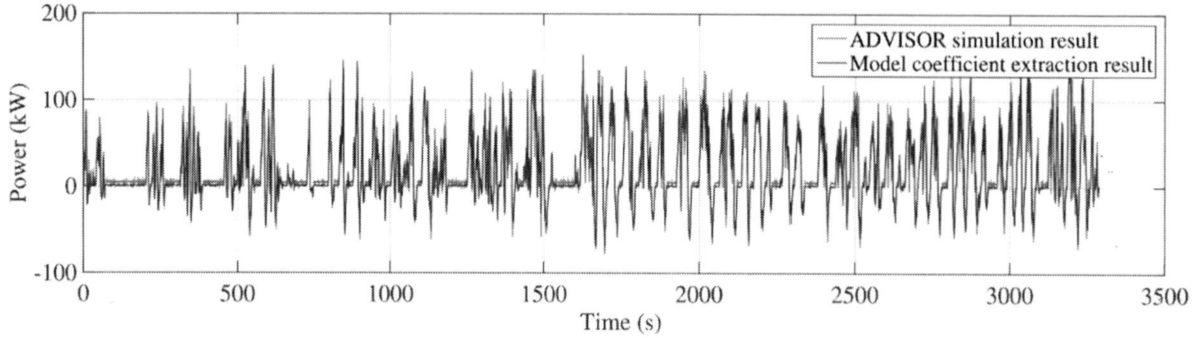

Figure 8. Powertrain model validation result.

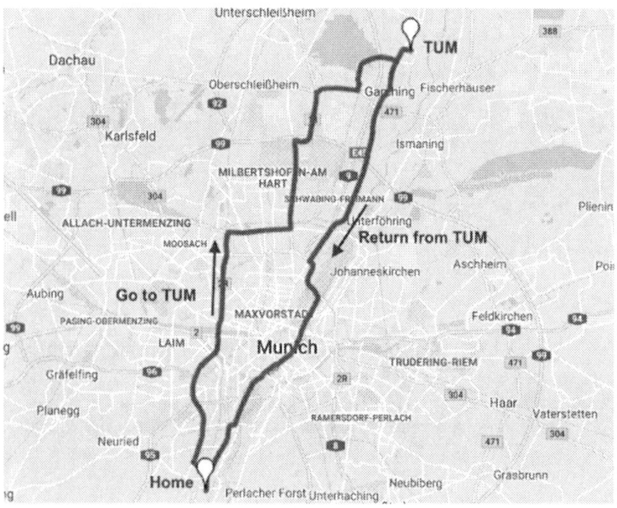

Figure 9. Routes from home to TUM and vice versa.

TABLE II. A ROUTE INFORMATION.

Route	Distance (km)	Avg. slope (degree)	Avg. speed (km/h)
Home to TUM	31.2	-0.1728	24.67
TUM to Home	26.2	0.2221	22.24

D. Case study 1: Fast Driving Energy Estimation

We extract a route going to Technische Universitat Munchen (TUM) and another one returning from TUM, as shown in Figure 9. Table II summarize the information of the routes: distance, average slope along the route and average vehicle speed.

The simulation results for the routes are shown in Figure 10. The first two graphs show the road altitude from home to TUM and the speed profile that we obtained from the road speed limit and the Google Maps traffic information. The third graph shows the corresponding power and energy consumption over time. The power consumption is low in the first half compared with the second half because the degree of negative slope along the road is high. Fourth and fifth graphs show the road altitude and speed profile from TUM to home. The sixth graph shows the power and energy consumption. The road slope is positive in this case. Therefore, the energy consumption is higher than the energy consumption to go to

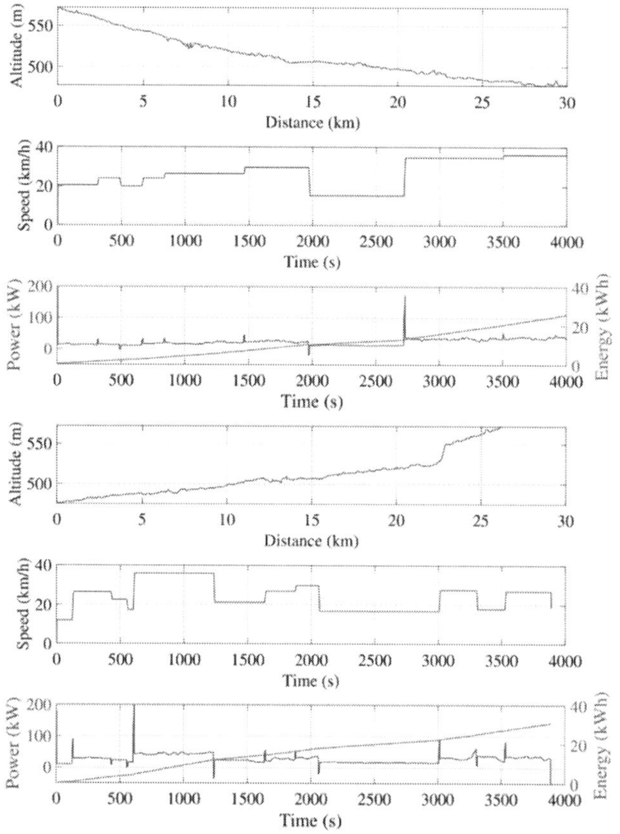

Figure 10. Simulation result from home to campus (first to third graphs) and from the campus to home (fourth to sixth graphs).

TUM. The driving distance going to TUM is longer than the route returning from TUM. However, the energy consumption to go to TUM (27.4 kWh) is nearly 10% lower than the energy consumption to return from TUM (30.8 kWh) because of the road slope. The proposed equation-form energy model helps us immediately estimate the energy consumption along the road, taking slope and traffic into consideration, and to decide which route is economic based on the fast simulation results.

E. Case Study 2: Battery Size Analysis

One important merit of the proposed power model is to estimate energy consumption in a short time, which is useful for iteration-based parameter sizing. For example, we can find the optimal battery size using iterative vehicle simulation. A short simulation time is mandatory in this approach. We perform the driving simulation on a flat 100 km distance with different battery pack sizes. We assume that the vehicle speed is 50 km/h, which is the average bus speed in the suburb of the city. We assume that the battery pack voltage is the same, and additional battery modules are attached in parallel. We assume that the battery pack is ideally balanced during battery charging and discharging.

We performed the vehicle simulations by changing the battery size from 70% of nominal battery size of BYD K9 to to 130%. Figure 11 shows the simulation results. As we increase battery size, the driving range also increases. However, the driving efficiency (energy consumption per unit distance) decreases because of the increase in battery weight. The driving range increases nearly 28% if we increase the battery size by 30%. On the other hand, the driving efficiency decreases up to 21%. Therefore, we should carefully decide the battery size with the consideration of cost of electricity per unit distance (efficiency) and bus service time (driving range with a fully charged battery).

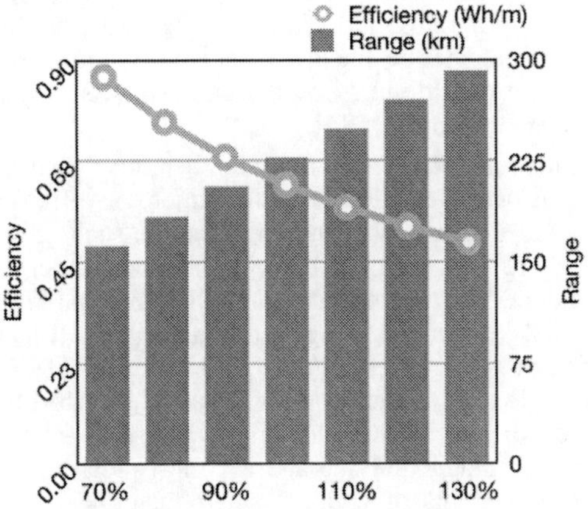

Figure 11. Simulation results by battery size.

V. CONCLUSION

We have proposed an improved EV range estimator incorporating (i) parameter extraction for complex EV simulation, (ii) equation-form powertrain modeling by coefficient extraction and (iii) fast vehicle energy simulation with a traffic- and altitude-aware driving cycle. We introduced two case studies as application of the proposed range estimator: (i) fast energy consumption estimation along the route information and (ii) bus battery sizing considering driving efficiency and range.

The estimator can work either offline, by estimating the range upfront without intermediate updates like a traditional GPS navigator, or online, refining the estimate at the cost of a route re-calculation. Our range estimator is meant as a "plug-in" for traditional or traffic-aware (e.g., Google Maps) GPS navigators, allowing route decisions besides traditional information based on travel time and route distance.

ACKNOWLEDGMENT

This work was supported by the National Research Foundation of Korea (NRF) grant funded by the Korean government (MSIT) (No. 2020R1F1A1074843).

REFERENCES

[1] European Automobile Manufacturers Association (ACEA), "Making the transition to zero-emission mobility," [Online]Available: https://www.acea.be/publications, June 2018.

[2] J. Wang, I. Besselink, and H. Nijmeijer, "Battery electric vehicle energy consumption prediction for a trip based on route information," Proceedings of the Institution of Mechanical Engineers, Part D: Journal of Automobile Engineering, vol. 232, no. 11, 2017, pp. 1528–1542.

[3] Y. Zhang, W. Wang, Y. Kobayashi, and K. Shirai, "Remaining driving range estimation of electric vehicle," in 2012 IEEE International Electric Vehicle Conference (IEVC). IEEE, 2012, pp. 1–7.

[4] "BYD Product," BYD Europe Official Website, [Online] Available: https://www.bydeurope.com [accessed: 2020-05-01].

[5] A. M. Andwari et al., "A review of battery electric vehicle technology and readiness levels," Renewable and Sustainable Energy Reviews, vol. 78, 2017, pp. 414–430.

[6] R. Zhang et al., "State of the art of lithium-ion battery SOC estimation for electrical vehicles," Energies, vol. 11, no. 7, 2018, pp. 1–36.

[7] Y. Wang et al., "Research on driving range estimation for EVs based on corrected battery model," in SAE World Congress & Exhibition, 2015.

[8] A. Bocca, A. Macii, E. Macii, and M. Poncino, "Composable battery model templates based on manufacturers' data," IEEE Design & Test, vol. 35, no. 3, 2018, pp. 66–72.

[9] M. Uitz et al., "Aging of Tesla's 18650 lithium-ion cells: Correlating solid-electrolyte-interphase evolution with fading in capacity and power," Journal of The Electrochemical Society, vol. 164, no. 14, 2017, pp. A3503–A3510.

[10] S. Kachroudi, M. Grossard, and N. Abroug, "Predictive driving guidance of full electric vehicles using particle swarm optimization," IEEE Trans. on Vehicular Technology, vol. 61, no. 9, 2012, pp. 3909–3919.

[11] K. Vatanparvar, J. Wan, and M. A. Al Faruque, "Battery-aware energy-optimal electric vehicle driving management," in International Symposium on Low Power Design (ISLPED), 2015, pp. 353–358.

[12] D. Baek and N. Chang, "Runtime power management of battery electric vehicles for extended range with consideration of driving time," IEEE Trans. on Very Large Scale Integration System, 2019, pp. 549–559.

[13] X. Yuan, C. Zhang, G. Hong, X. Huang, and L. Li, "Method for evaluating the real-world driving energy consumptions of electric vehicles," Energy, vol. 141, 2017, pp. 1955–1968.

[14] A. Gonzalez-Castellanos, D. Pozo, and A. Bischi, "A non-ideal linear operation model for a li-ion battery," arXiv:1901.04260, 2019.

[15] G. Liu, M. Ouyang, L. Lu, J. Li, and J. Hua, "A highly accurate predictive-adaptive method for lithium-ion battery remaining discharge energy prediction in electric vehicle applications," Applied Energy, vol. 149, 2015, pp. 297–314.

[16] Y. Chen et al., "A systemc-ams framework for the design and simulation of energy management in electric vehicles," IEEE Access, vol. 7, 2019, pp. 25 779–25 791.

[17] D. Baek et al., "Battery-aware operation range estimation for terrestrial and aerial electric vehicles," IEEE Transactions on Vehicular Technology, 2019, pp. 5471–5482.

[18] T. Markel, A. Brooker, T. Hendricks, V. Johnson, and K. Kelly, "Advisor: a systems analysis tool for advanced vehicle modeling," Journal of Power Sources, 2002.

[19] Y. Chen, E. Macii, and M. Poncino, "A circuit-equivalent battery model accounting for the dependency on load frequency," in Proceedings of the Conference on Design, Automation & Test in Europe. European Design and Automation Association, 2017, pp. 1177–1182.

[20] Y. Chen, D. J. Pagliari, E. Macii, and M. Poncino, "Battery-aware design exploration of scheduling policies for multi-sensor devices," in ACM/IEEE Great Lakes symposium on VLSI (GLSVLSI), 2018.

[21] Google Maps, [Online] Available: https://www.google.com/maps [accessed: 2020-05-01].

[22] "BYD K9," Land Transport Guru, [Online] Available: https://landtransportguru.net/byd-k9/ [accessed: 2020-05-01].

[23] K. Norregaard, B. Johnsen, and C. H. Gravesen, "Battery Degradation in Electric Buses," Danish Technological Institute, Tech. Rep., 2016.

[24] "Federal Transit Bus Test," The Larson Institute, Tech. Rep., 2014.

When Privacy Protection Meets Non-Intrusive Load Monitoring: Trade-off Analysis and Privacy Schemes via Residential Energy Storage

Sangyoung Park

Smart Mobility Systems
Technische Universität Berlin
Einstein Center Digital Future
Berlin, Germany
Email: sangyoung.park@tu-berlin

Andrea Cominola

Smart Water Networks
Technische Universität Berlin
Einstein Center Digital Future
Berlin, Germany
Email: andrea.cominola@tu-berlin.de

***Abstract*—Non-Intrusive Load Monitoring (NILM) algorithms are actively being researched to disaggregate the electricity usage of a whole household into the contribution of individual appliances. While understanding the usage patterns of individual appliances can be beneficial for flattening the peak demand, reducing the electricity bill, and improving the energy usage efficiency, NILM algorithms raise privacy concerns. Residential energy storage could be used to relieve such concerns by modifying the monitored electricity profile. However, residential energy storage systems are yet costly, and hence assessing the financial overhead of privacy protection techniques is important. In this paper, we provide motivational examples and early results on how much residential energy storage would be required to fool a state-of-the-art NILM algorithm. Our preliminary results on the trade-off between NILM accuracy and privacy protection indicate that some intuitive approaches that require a significant amount of battery capacity are not necessarily the most effective in reducing the disaggregation accuracy.**

Keywords–Privacy; Non-Intrusive Load Monitoring; smart meter; energy storage.

I. INTRODUCTION

Smart meters are becoming an essential component in smart grids because they allow high-resolution, real-time monitoring of electricity generation/consumption and communication of the pricing information [1]. Smart meter information is key for utility companies to provide information for the design and implementation of demand side management strategies and better cope with fluctuating electricity demands [2]. This results in more effective matching of the electricity demand and generation. In addition, smart meter information is expected to also bring benefits to electricity consumers, as it provides them with more transparent billing information and enables a better control over their electricity usage [3].

Motivated by the need to provide demand management programs with better understanding, models, and forecasts of electricity demand with high spatial and temporal resolution, there have been recent rapid advances in the development of Non-Intrusive Load Monitoring (NILM) algorithms. NILM algorithms allow automatic decomposition of the aggregated electric load measured by a smart meter at the household level into the electricity usage patterns of individual appliances (e.g., fridge, air conditioning system, etc.) [4]. One digital AC monitor, i.e., a smart meter, measuring the single-phase power into a household instead of individual sensors for each appliance, suffices to provide a low cost and non-intrusive solution. The information on electricity demand obtained via NILM algorithms can be used to support consumption-based feedback programs and customer segmentation for demand management, flatten the peak demand, identify faulty appliances, and provide hints on the effectiveness of demand side management programs [5].

However, NILM algorithms inherently generate privacy concerns. Besides detailed information on electricity usage behaviors, sensitive private information such as how many people are present in a home at a given time, absence of a resident, or even gender and age information could be potentially estimated using NILM algorithms [6][7]. What is even more of a concern is that, from the user side, it would be impossible to detect whether or not a NILM algorithm is being used by the utility provider or a third-party unless explicitly communicated. Resolving the existing trade-off between the expected benefits of coupled smart metering-NILM systems and the privacy challenges that they include is thus key to facilitate the development of privacy-aware smart meter deployments.

A potentially effective way of alleviating the privacy concerns is to physically modify the electricity profile seen by the smart meters [8]. This can be achieved by the use of a residential energy storage. Existing commercial products, such as Tesla Powerwall [9] or Encharge from Enphase [10], are primarily designed for the purpose of compensating the fluctuating power generation of the rooftop solar arrays. However, their marketability is still being carefully evaluated as it is highly dependent on the battery purchase cost, depreciation cost due to aging as well as the local electricity cost. Hence, using a residential energy storage for privacy protection could harm the return-on-investment unless it is properly sized and utilized.

In this paper, we investigate the cost-effectiveness of the potential usage of a residential energy storage for privacy protection against NILM algorithms. We first perturb the electricity signal seen by the single-point smart meter of a sample household to simulate different privacy protection scenarios. These scenarios include the superimposition of simple Gaussian noise on the smart meter signal, or a flattened signal due to the usage of residential energy storage (i.e., batteries) of different size and cost. We then investigate the trade-off between privacy protection and NILM capabilities by analyzing how the performance of a state-of-the-art NILM algorithm changes in scenarios determined by increasing privacy protection. The contributions of this paper can be summarized as follows.

- We analyze the accuracy and effectiveness of the NILM algorithm over modified household consumption profiles using residential energy storage (i.e., batteries).
- We analyze the trade-off between different privacy protection schemes and the battery size needed to implement them.
- We provide recommendations on how to effectively use residential energy storage to protect the privacy of electricity users against NILM algorithms.

The ultimate goal of this paper is not to condemn the detailed investigation of household energy demand at the end use level, which can be very important to provide demand management programs, or also coordinated water-energy conservation/peak shifting programs. Rather, we aim to foster the conversation about the privacy implications of NILM and to set the basis for a wider discussion about the conflicting trade-off of demand management programs in terms of technical feasibility, overall economic and environmental benefits and social acceptance.

The rest of the paper is organized as follows. Section II summarizes the related works on the usage of an energy storage in the residential sector and NILM. Section III describes the system setup and experimental settings for the NILM algorithm used in this study. Section V analyzes some early experimental results. Section VI concludes the paper.

II. Related Works

The development of smart energy grids and the uptake of smart metering devices have fostered many recent research efforts in two potentially conflicting directions, i.e., (i) electricity demand modeling at the end use level, and (ii) privacy protection.

On the one hand, the increased availability of smart meter data at the household level and recorded with sub-daily sampling resolution has fostered the development of NILM algorithms for electricity demand modelling at the end use level, following the 1992 seminal study by Hart [11]. NILM algorithms thus aim at estimating the electricity consumption (or consumption pattern) of each appliance contributing to the aggregate electricity profile recorded by a single-point smart meter, installed at the household level. Given the aggregate household electricity consumption Y_t caused by N appliances and recorded by the single-point smart meter at time t, NILM algorithms estimate the non-metered consumption y_t^i of each individual appliance i, where

$$Y_t = \sum_{i=1}^{N} y_t^i + e_t \tag{1}$$

and e_t is the measurement noise. This process is non-intrusive, as the installation of multiple sensors at the appliance level is avoided. In the last two decades, several NILM algorithms have been proposed in the literature: *optimization-based* methods (e.g., integer or sparse optimization [12]); *pattern recognition* methods, which model the temporal structure of electricity signals (e.g., methods based on Markov Models [13]); *hybrid* methods [14]; and *unsupervised* algorithms [15]. More recent approaches primarily exploited deep neural networks (e.g., [4]) and explored the potential of transfer learning to generalize and transfer NILM algorithms among different domains [16]. The above methods have been extensively tested and cross-compared on benchmark data sets [17][18] and with widely adopted performance metrics [19].

On the other hand, the state-of-the-art literature includes approaches for *privacy protection* that use a residential energy storage to modify the usage pattern of appliances and human activities [8][20]. These studies often target flattening out, i.e., water-filling, the household electricity profile for privacy protection. However, this usually results in an extensive use of the storage requiring a large capacity and leads to accelerated aging, thus rendering the solution impractical. An alternative approach to privacy protection consists of randomizing the household electricity profile by adding, for instance, Gaussian noise [21]. However, the lack of a precise definition of *privacy* has limited so far the possibility to come up with a cost effective and general solution. Most previous works made use of the concept of signal "flatness" to quantify the privacy level of a modified electricity signal, and formulate it by proxy indicators, including the sum of Root Mean Square (RMS) error, the mutual information [22], or the entropy [23] between the original and the modified profile. However, such metrics are only weakly linked to the people's perception of privacy as it is hard to grasp the feeling of how much private information can be extracted by just flattening the electricity profile.

In this paper, we aim neither at introducing a comprehensive definition of privacy in the residential electricity sector, nor at proposing new NILM algorithms. Rather, here, we explore the influence of different privacy schemes on the accuracy of state-of-the-art NILM algorithms. The different privacy schemes are generated by perturbation of the household electricity signal as caused by operation of a residential energy storage device. Thus, we measure privacy protection as the ability of a residential energy storage to alter the household electricity signal seen by the smart meter and hamper accurate identification of end use electricity consumption via NILM algorithms. Such an approach entails that trade-offs between the privacy scheme defined by the usage (and thus size and cost) of the residential energy storage and the accuracy may emerge. The ultimate goal of this paper is, thus, to analyze this trade-off and come up with recommendations to foster the identification of cost-effective solutions for privacy protection in the residential electricity sector, as well as contributing to the overall discussion on the benefits and challenges of the advanced metering and analytics tools characterizing the ongoing digitalization of the utility and house sectors.

III. Material and Methods

A. Residential Electricity System

Figure 1 shows the system setup of the sample household considered in this study. A smart meter monitors the whole household energy consumption and communicates the value to the utility provider. The utility provider makes use of the smart meter information for demand side management. The utility provider could run NILM algorithms with the consent of the residents or there could be a third-party acting on its own initiative, potentially malicious, running NILM algorithms. In the household, there are a number of electricity appliances, such as washing machines, TVs, ovens, heat pumps, HVAC, dryers, etc. Optionally, the house includes a battery storage at home connected to the low-voltage AC network through a bidirectional DC/AC converter. The storage resembles products like Powerwall from Tesla (13.5 kWh) [9] or Encharge from Enphase (3.5 kWh and 10.5 kWh) [10]. There could be multiple purposes of installing the storage such as leveraging the electricity price difference over time, stabilizing the local low distribution grid by participating as a primary control reserve, etc. In this work, we exploit the presence of a battery storage for its additional function of privacy protection, as it can be used to modify the electricity signal seen by the smart meter, thus hiding the actual electricity usage of individual appliances.

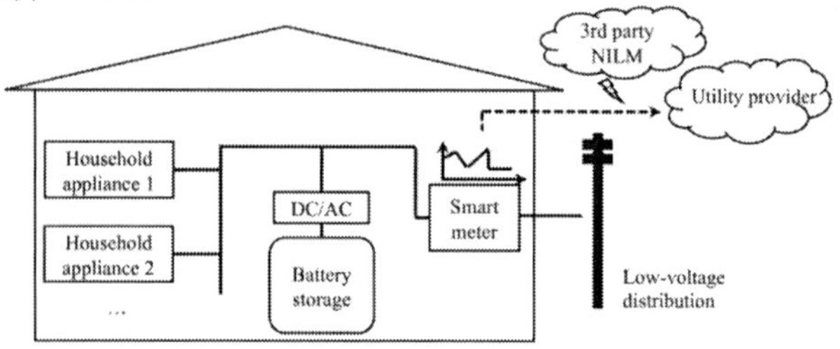

Figure 1. Electricity system setup for the sample household.

B. *Sparse NILM algorithm*

In this paper, we rely on state-of-the-art NILM algorithms to explore the influence of different privacy schemes via residential energy storage on NILM capabilities.

Specifically, we adapt the open source NILM algorithm proposed by [13], a computationally efficient algorithm for online real-time NILM based on a super-state Hidden Markov Model (HMM) and a modified Viterbi algorithm that efficiently manages the sparsity in HMM matrices (the original code used in [13] is available on GitHub at [24]). In short, the algorithm follows the following sequential process. First, the data sub-metered at the level of individual appliance are analyzed to create probability mass functions for each appliance and identify its load states. Second, a super-state HMM is created by combining the individual load states identified in the previous step. This builds the actual NILM model. Finally, the super-state HMM model is combined with a sparse Viterbi algorithm to perform NILM and disaggregate the observed aggregate smart meter load into the estimated contribution of each appliance. The Viterbi algorithm is a dynamic programming algorithm that is usually used to estimate the most likely sequence of hidden states associated with a measured output of a process modelled with HMM. The modified Viterbi algorithm proposed in [13] exploits the high rate of zero-probability terms in the HMM matrices, thus avoiding unnecessary computations and efficiently returning a solution. The resulting sparsity-based NILM algorithm has been demonstrated to outperform other state-of-the-art NILM algorithms. Moreover, it can handle disaggregation with data sampled with low-frequency (e.g., 1 min), while many other algorithms require higher frequencies, and is scalable, i.e., can handle a large number of super-states [13].

We assess the NILM performance of the sparsity-based NILM algorithm under different privacy-protection scenarios according to two performance metrics, i.e., (i) the *Finite-State F-score* (FS-fscore) and (ii) the *Mean Absolute Percentage Error* (MAPE). The FS-fscore was first introduced in [19] as an alternative to conventional F-score to account for inaccuracies in non-binary classification and is formulated as follows:

$$FS_i = \frac{2 \times PC_i \times RC_i}{PC_i + RC_i} \tag{2}$$

where RC_i and PC_i are the *recall* and *precision*, respectively, formulated for each appliance i in order to take into account the classification inaccuracy $RC_i = \frac{TP_i - inacc_i}{TP_i + FN_i}$ and $PC_i = \frac{TP_i - inacc_i}{TP_i + FP_i}$. The $inacc$ term can be defined, for each appliance i, as:

$$inacc = \sum_{t=1}^{H} \frac{|\hat{x}_t^i - x_t^i|}{K^i} \tag{3}$$

where $\hat{x}_t^i$ and x_t^i are the estimated and observed states of appliance i at time t, H is the considered time horizon, K^i is the number of states of appliance i, TP_i, FP_i, and FN_i are the number of true positive, false positive, and false negative events, respectively. Overall, the FS-fscore metric evaluates how good a NILM algorithm is in classifying the operating states of the considered appliances. MAPE is formulated as follows, for appliance i:

$$MAPE = \frac{1}{H} \sum_{t=1}^{H} \left| \frac{y_t^i - \hat{y}_t^i}{y_t^i} \right| \tag{4}$$

where the notation is consistent with the variables previously defined in this paper. After calculating the FS-fscore and MAPE performance metrics for each appliance, we compute their average value across appliances to assess how affected the NILM performance is on average, for each privacy protection scheme.

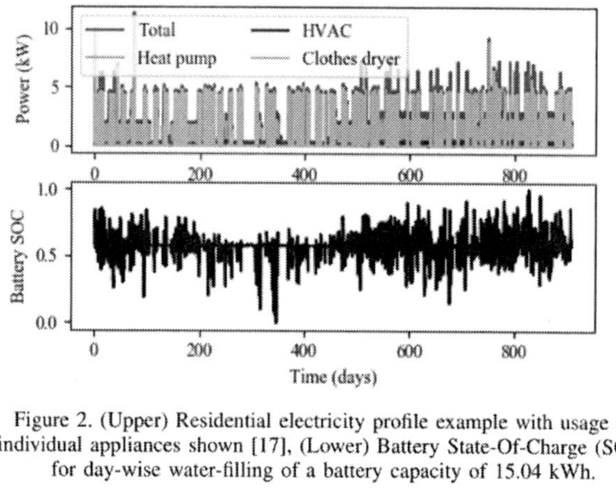

Figure 2. (Upper) Residential electricity profile example with usage of individual appliances shown [17], (Lower) Battery State-Of-Charge (SOC) for day-wise water-filling of a battery capacity of 15.04 kWh.

C. *AMPds Dataset*

To simulate the electricity consumption of the different appliances included in the sample household of Figure 1 and train the sparse NILM algorithm, we used the appliance-level data included in the AMPds dataset [17]. The AMPds dataset contains electricity, water, and natural gas measurements at one minute intervals for 2 years of monitoring. The data was collected at a house in Canada, where a family of three live. In this study, we considered the original version of the dataset, which includes data for 19 appliances monitored over a period of 1 year [25]. Electricity was measured with 1 Hz frequency, and 21 loads including a furnace, a fridge, a heat pump, a clothes dryer, TVs, a wall oven, etc., which makes it suitable for developing and evaluating NILM algorithms. In the trade-off analysis presented in this work, we considered an increasing number of appliances in the AMPds - from 3 to 10 - selected based on their ascending total electricity usage over the 1-year time series.

IV. Privacy Protection Schemes

One way to protect privacy is to completely flatten out the usage profile like water-filling strategy. However, water-filling strategy could be prohibitive due to the high usage of battery. For example, to completely filter out the whole house energy consumption reported in the AMPds dataset shown in Figure 2 (top) over a year, a battery of usable capacity 15.04 kWh has to be used. Assuming a battery system cost of 700 USD/kWh including the pack and the electronic equipment, this would result in more than 10,000 USD for purchasing the battery system. Assuming a lifetime of 10 years, which is equivalent to the warranty period of Tesla Powerwall, this would roughly equate to 1,000 USD/year solely for protecting privacy. There are other factors that add to cost, such as sub-optimal exploitation of Time-Of-Use (TOU) tariffs and accelerated aging due to repeated cycle charge/discharge. Such factors, however, will be left as future works, and we focus on the required capacity in this work. Next, we present various methods to modify the original electricity consumption profile to protect the privacy against the Sparse NILM algorithm presented in Section III-B.

M1: Add Gaussian noise to the whole duration of the profile. The method is intended to make the profile hard to analyze regardless of the time of use and appliance usage patterns. We control the intensity of the noise to see the impact. Larger noise leads to reduced accuracy of the NILM algorithms. However, it also involves more usage of the battery capacity. The upper graph in Figure 3 shows such an example. We have introduced a Gaussian noise with a standard deviation of 3 A. The mean of the noise over the long term is of course zero, but temporary increase and decrease in cumulative energy from the battery is inevitable, as can be seen from Figure 3. We observe that a naive implementation of M1 requires an unnecessarily large battery capacity, so we periodically reset the SOC of the battery to a pre-defined level. The usage pattern can still be modified while efficiently making use of the battery capacity.

M2: Add Gaussian noise only when a particular appliance of interest is used. For M2, we add noise only to time slots where a particular appliance is being used. We test this scheme in order to find out whether we can selectively hide the usage profile of an appliance, which will be using less battery capacity compared with M1.

M3: Water-filling for a particular appliance of interest. For M3, we see whether hiding the usage pattern by flattening out the appliance profile while it is being used is effective in reducing the accuracy of the NILM algorithm. Figure 4 shows how it modifies the profile. In the modified profile, the load looks rectangular rather than preserving all the "shapes" of the consumption. Because the load is averaged, the peak becomes lower than the original profile.

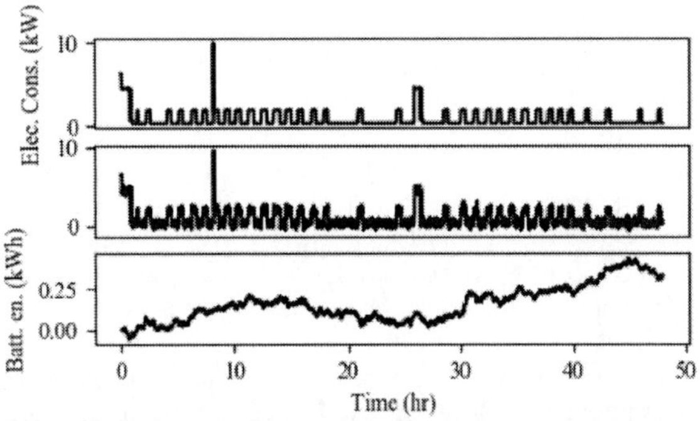

Figure 3. Original (upper) and modified (middle) whole household electricity consumption using privacy protection scheme M1, and the corresponding battery energy level (lower) over two days.

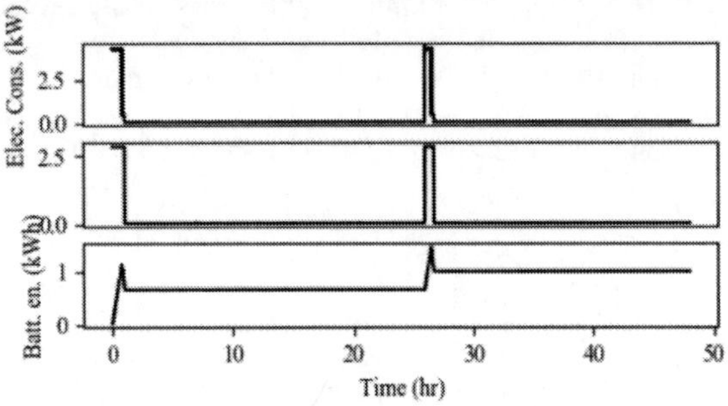

Figure 4. Original (upper) and modified (middle) profile of the clothes dryer using privacy protection scheme M3, and corresponding battery profile (lower) over two days.

M4: Spread-out the electricity consumption of a particular application. What can be seen from Figure 4 is that the appliance, which draws significant current, is still very visible to the human eye in the overall profile. Hence, we also attempt to modify the peak value with the support of a battery in varying degrees. Therefore, we explore M4, which reduces the size of the peak and spreads out the electricity consumption evenly among all time slots.

M5: Erase an appliance's consumption. This is similar to M4, but completely erases an appliance's usage profile from the whole household energy consumption by charging the battery storage. An equivalent amount of energy is discharged over the whole period of time just like M4.

M6: Day-wise water-filling for the whole electricity consumption profile. Finally, M6 performs a water-filling technique, which results in a flat profile seen by the smart meter. However, doing this for a whole year, i.e., flat profile for a year, is prohibitive because of the seasonal variations in electricity consumption, which mandates the use of an excessively large capacity battery, i.e., tens of kWh of battery capacity over the course of two years just for introducing the Gaussian noise of 6 A standard deviation. Therefore, we investigate day-wise water-filling, which means the smart meter will see a flat profile within a day, but varying electricity profile among days.

We evaluate the reduction in disaggregation accuracy and the required battery capacity, and hence the cost, for each method to provide an insight into developing cost-effective techniques.

V. EXPERIMENTAL RESULTS

In this section, we present the FS-fscore and MAPE obtained for NILM under the different privacy protection schemes detailed in the previous section. The overall evaluation procedure is shown in Figure 5. For the sake of analysis, we have preprocessed the AMPds dataset and prepared 8 sets of the Whole Household Energy (WHE) consumption profile. Each set reconstructs the WHE profile using the top 3, 4, ···, 10 appliances, sorted by increasing contribution to the whole household energy consumption. Generally speaking, the disaggregation accuracy is expected to decrease when there are more appliances to disaggregate. Hence, we varied the number of appliances to assess the effectiveness of the NILM algorithm in all cases. We build the NILM models from the preprocessed WHE profiles. Then, the profile modification algorithms from Section IV are applied to simulate different privacy protection schemes. The modified profiles serve as inputs to the sparse Viterbi algorithm, which estimates the most likely sequence of hidden states (i.e., appliance states). Finally, the NILM performance is assessed by FS-fscore and the MAPE metrics.

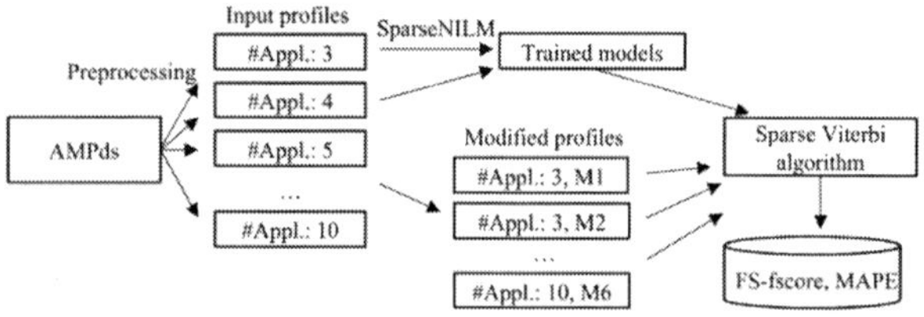

Figure 5. Overall evaluation procedure of NILM under different privacy protection schemes.

Figure 6 shows the FS-fscore and MAPE averaged across all appliances versus the required battery capacity of the modification algorithm. The required battery capacity varies

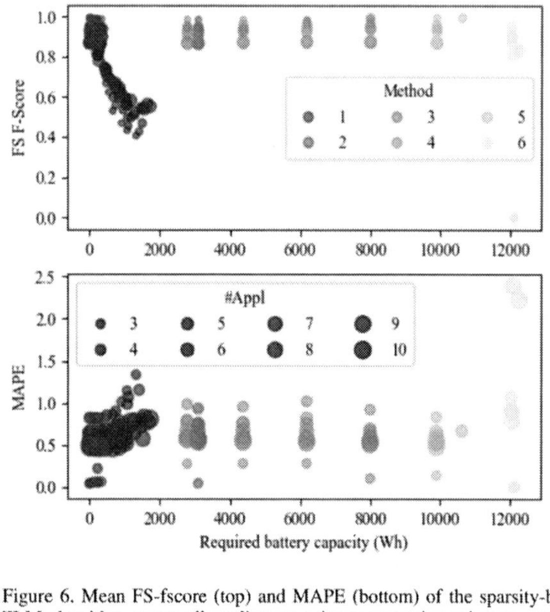

Figure 6. Mean FS-fscore (top) and MAPE (bottom) of the sparsity-based NILM algorithm across all appliances, privacy protection schemes, and their required battery capacity.

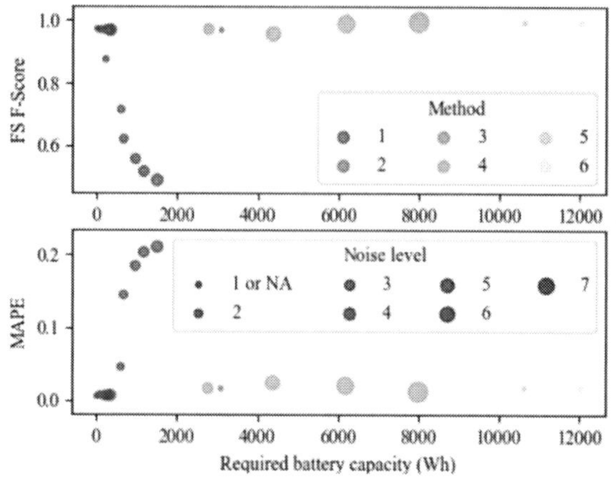

Figure 7. FS-fscore (top) and MAPE (bottom) of the cloths dryer (appliance of interest) vs required battery capacity of the profile modification methods. The number of appliances is 5 for all the results. Marker color varies for each profile modification method (M1-M6). Marker size is proportional to the noise level of each method (the higher, the more noisy).

from 0 kWh to 12 kWh depending on the six modification methods. M6 requires the most amount of battery capacity as it aims at a completely flat profile. The sizes of markers correspond to the number of appliances and the colors of markers correspond to the privacy protection scheme. It can be seen that, overall, a larger number of appliances leads to slightly less accurate disaggregation within each method. The privacy protection schemes M1 and M2 require the least amount of battery capacity as Gaussian noise is relatively small and averages out over a long period of time, as can be seen from Figure 2. The most noticeable point is that both FS-fscore and MAPE do not dramatically change for methods M3 to M6.This might be due to the fact that the trained NILM algorithm can still provide accurate estimates for the status of most of those appliances that operate in a very limited range, or that fall within the range seen by the smart meter, and thus the average performance is still good in terms of FS-fscore. However, the majority of MAPE values are between 0.5 and 1.2, indicating that the different privacy protection algorithms can successfully hamper the identification of detailed appliance profiles. Overall, the two most cost-effective methods to protect privacy against the considered sparsity-based NILM algorithm appear to be, in this specific sample case, M1 and M2. The more battery capacity is utilized for M1 and M2, the worse FS-fscore and MAPE become. The FS-fscore drops almost to 0.4 when a battery capacity of around 2 kWh is used for obfuscating the original profile. Other methods require much more battery capacity, but they do not necessarily result in a larger drop in NILM accuracy, as far as the FS-fscore is concerned.

Figure 7 shows the FS-fscore and MAPE for a particular appliance of interest, in this case, the clothes dryer, instead of the mean of the metrics across all appliances. For the sake of illustration, only the results obtained for the NILM experiments including 5 appliances are reported. The marker size corresponds to the magnitude of profile modifications, in each privacy protection scheme. For M1 and M2, it is the standard deviation of the Gaussian noise. For M4, it is the amount of change in the mean of the appliance. For those methods not requiring different profile modification magnitude, the marker sizes are depicted as the smallest. In this graph, it is clearly visible that the disaggregation accuracy is lower for larger Gaussian noise with M1. Surprisingly, M5 and M6, which completely hide an appliance's profile, do not have a profound impact on the metrics of this appliance. This might be due to the dominance of the clothes dryer profile over other appliances. When Gaussian noise is imposed on the aggregate signal, this specific profile becomes harder to identify, but, in the other cases, a guess from the sparsity-based NILM algorithm for this specific appliance is still accurate. Another noticeable thing is that M2, which is dedicated to modifying the appliance's record, has nearly no impact on the metrics, while M1 degrades the performance of the sparsity-based NILM algorithm the most. Such an observation motivates further systematic research on the topic, including a sensitivity analysis on how these results change for different NILM algorithms and with different performance metrics.

VI. Conclusion and Future Work

In this paper, we investigated the effectiveness of a number of heuristic algorithms making use of a residential battery storage in preserving privacy against a NILM algorithm. Most of the prior works are based on the water-filling technique, i.e., flattening out the consumption profile as much as possible. However, they require a significant amount of battery storage while the amount of privacy protection is hard to quantify. Therefore, we focus on the accuracy metrics of a NILM algorithm and quantify the effectiveness of the profile modification methods. Our preliminary results indicate that some intuitive methods do not necessarily lead to a significant reduction in the disaggregation performance of the NILM algorithm, even though they require significant battery capacity. This points towards further systematic research tailored to providing protection against NILM algorithms while minimizing the battery cost overhead.

Also, the cost analysis in this paper is restricted to battery capacity. A holistic cost analysis incorporating the electricity bills under flexible pricing policy, the interplay with the rooftop solar arrays (when available), and the impacts of different battery usage schemes on battery aging rates should be performed. Finally, future research should look at better linking the battery capacity to the specific appliance (or appliances) that a customer would like to hide for privacy protection, if some are more privacy-sensitive than others.

Acknowledgment

The authors acknowledge the support of the Einstein Center Digital Future (Berlin, Germany).

References

[1] F. Tounquet and C. Alaton, "Benchmarking smart metering deployment in the EU-28," European Commission, Final Report, 2020.

[2] A. Mohsenian-Rad, V. W. S. Wong, J. Jatskevich, R. Schober, and A. Leon-Garcia, "Autonomous demand-side management based on game-theoretic energy consumption scheduling for the future smart grid," *IEEE Transactions on Smart Grid*, vol. 1, no. 3, pp. 320–331, 2010.

[3] S. S. S. R. Depuru, L. Wang, V. Devabhaktuni, and N. Gudi, "Smart meters for power grid — challenges, issues, advantages and status," in *IEEE/PES Power Systems Conference and Exposition*, 2011, pp. 1–7.

[4] J. Kelly and W. Knottenbelt, "Neural nilm: Deep neural networks applied to energy disaggregation," in *ACM International Conference on Embedded Systems for Energy-Efficient Built Environments*, 2015.

[5] H. Cao, C. Beckel, and T. Staake, "Are domestic load profiles stable over time? an attempt to identify target households for demand side management campaigns," in *IEEE Industrial Electronics Society Conference (IECON)*, 2013, pp. 4733–4738.

[6] A. Molina-Markham, P. Shenoy, K. Fu, E. Cecchet, and D. Irwin, "Private memoirs of a smart meter," in *ACM Workshop on Embedded Sensing Systems for Energy-Efficiency in Building (BuildSys)*, 2010.

[7] P. McDaniel and S. McLaughlin, "Security and privacy challenges in the smart grid," *IEEE Security Privacy*, vol. 7, no. 3, pp. 75–77, 2009.

[8] L. Yang, X. Chen, J. Zhang, and H. V. Poor, "Cost-effective and privacy-preserving energy management for smart meters," *IEEE Transactions on Smart Grid*, vol. 6, no. 1, pp. 486–495, 2015.

[9] "Powerwall — Tesla," 2020, URL: https://www.tesla.com/powerwall [accessed: 2020-07-09].

[10] "Install Solar Backup Battery - Ensemble^{TM} Technology — Enphase," 2020, URL: https://enphase.com/en-us/ensemble-technology-enphase-installers [accessed: 2020-07-09].

[11] G. W. Hart, "Nonintrusive appliance load monitoring," *Proceedings of the IEEE*, vol. 80, no. 12, pp. 1870–1891, 1992.

[12] D. Piga, A. Cominola, M. Giuliani, A. Castelletti, and A. E. Rizzoli, "Sparse optimization for automated energy end use disaggregation," *IEEE Transactions on Control Systems Technology*, vol. 24, no. 3, pp. 1044–1051, 2015.

[13] S. Makonin, F. Popowich, I. V. Bajić, B. Gill, and L. Bartram, "Exploiting HMM sparsity to perform online real-time nonintrusive load monitoring," *IEEE Transactions on Smart Grid*, vol. 7, no. 6, pp. 2575–2585, 2016.

[14] A. Cominola, M. Giuliani, D. Piga, A. Castelletti, and A. E. Rizzoli, "A hybrid signature-based iterative disaggregation algorithm for non-intrusive load monitoring," *Applied Energy*, vol. 185, pp. 331–344, 2017.

[15] R. Bonfigli, S. Squartini, M. Fagiani, and F. Piazza, "Unsupervised algorithms for non-intrusive load monitoring: An up-to-date overview," in *2015 IEEE 15th International Conference on Environment and Electrical Engineering (EEEIC)*. IEEE, 2015, pp. 1175–1180.

[16] M. D'Incecco, S. Squartini, and M. Zhong, "Transfer learning for non-intrusive load monitoring," *IEEE Transactions on Smart Grid*, vol. 11, no. 2, pp. 1419–1429, 2019.

[17] S. Makonin, B. Ellert, I. V. Bajic, and F. Popowich, "Electricity, water, and natural gas consumption of a residential house in Canada from 2012 to 2014," *Scientific Data*, vol. 3, no. 160037, pp. 1–12, 2016.

[18] J. Z. Kolter and M. J. Johnson, "Redd: A public data set for energy disaggregation research," in *Workshop on data mining applications in sustainability (SIGKDD), San Diego, CA*, vol. 25, no. Citeseer, 2011, pp. 59–62.

[19] S. Makonin and F. Popowich, "Nonintrusive load monitoring (NILM) performance evaluation," *Energy Efficiency*, vol. 8, no. 4, pp. 809–814, 2015.

[20] G. Kalogridis, Z. Fan, and S. Basutkar, "Affordable privacy for home smart meters," in *IEEE International Symposium on Parallel and Distributed Processing with Applications Workshops*, 2011, pp. 77–84.

[21] E. Liu and P. Cheng, "Achieving privacy protection using distributed load scheduling: A randomized approach," *IEEE Transactions on Smart Grid*, vol. 8, no. 5, pp. 2460–2473, 2017.

[22] A. Pröbstl, S. Park, S. Steinhorst, and S. Chakraborty, "Cost/privacy co-optimization in smart energy grids," in *Design, Automation Test in Europe Conference Exhibition (DATE)*, 2019, pp. 872–877.

[23] S. McLaughlin, P. McDaniel, and W. Aiello, "Protecting consumer privacy from electric load monitoring," in *Conference on Computer and Communications Security (CCS)*, 2011, pp. 87–98.

[24] S. Makonin, *The super-state hidden Markov model disaggregator that uses a sparse Viterbi algorithm for decoding.*, 2018 (accessed August 17, 2020). [Online]. Available: https://github.com/smakonin/SparseNILM

[25] S. Makonin, F. Popowich, L. Bartram, B. Gill, and I. V. Bajić, "Ampds: A public dataset for load disaggregation and eco-feedback research," in *2013 IEEE Electrical Power & Energy Conference*. IEEE, 2013, pp. 1–6.